Nano Dielectric Resonator Antennas for 5G Applications

Nano Dielectric Resonator Antennas for 5G Applications

Rajveer S. Yaduvanshi

Gaurav Varshney

CRC Press
Taylor & Francis Group
Boca Raton London New York

CRC Press is an imprint of the
Taylor & Francis Group, an **Informa** business

First edition published 2020
by CRC Press
6000 Broken Sound Parkway NW, Suite 300, Boca Raton, FL 33487-2742

and by CRC Press
2 Park Square, Milton Park, Abingdon, Oxon, OX14 4RN

© 2020 Taylor & Francis Group, LLC

CRC Press is an imprint of Taylor & Francis Group, LLC

Library of Congress Cataloging-in-Publication Data

Names: Yaduvanshi, Rajveer S., author. | Varshney, Gaurav, author.
Title: Nano dielectric resonator antennas for 5G applications / Rajveer S. Yaduvanshi, Gaurav Varshney.
Description: First edition. | Boca Raton, FL : CRC Press, 2020. | Includes bibliographical references and index.
Identifiers: LCCN 2020011030 (print) | LCCN 2020011031 (ebook) | ISBN 9780367465339 (hardback) | ISBN 9781003029342 (ebook)
Subjects: LCSH: Microwave antennas. | Antennas, Dipole. | 5G mobile communication systems--Equipment and supplies.
Classification: LCC TK7871.67.M53 Y33 2020 (print) | LCC TK7871.67.M53 (ebook) | DDC 621.382/4--dc23
LC record available at https://lccn.loc.gov/2020011030
LC ebook record available at https://lccn.loc.gov/2020011031

ISBN: 978-0-367-46533-9 (hbk)
ISBN: 978-1-003-02934-2 (ebk)

Typeset in Times
by Lumina Datamatics Limited

Contents

Preface

The nano dielectric resonator antenna operates at very small wavelengths, i.e., nanometers. Switching from microwave (1–100 GHz) to optical frequency can provide large bandwidth and compact size devices. They are different from patch antenna. DRA has more advantages and capabilities in terms of performance parameters. In DRA, microwave signal is applied to an input port matched to 50 ohms. The signal travels through microstrip line to slot. The microstrip line is a non-resonant transmission line which acts as an impedance transformer. And slot is resonant at operating frequency, which is dependent upon w/l (width to length ratio, known as aspect ratio). Slot can form short magnetic dipoles along its length, and E-fields current vector shall be normal to the slot surface. The operating frequency of microwave input and slot resonant frequency is generally kept the same for all antenna designs. Magnetic dipoles are generated due to slot structure and input excitations. The resonant modes are field pattern formed by magnetic or electric dipoles. Hence, slot can be represented by an equivalent parallel RLC circuit. These dipoles are coupled to DRA. A term loading factor is introduced by DRA. This is dependent upon coupling of magnetic dipoles from slot to DRA. The slot and DRA and microstrip lines are matched for 50 ohm impedance for better coupling. The quality factor of DRA has an important role for radiation. Resonant modes inside DRA are dependent upon DRA cavity nature and its structure. Reflections and refractions of microwave signals based on boundary conditions and material properties of cavity formed introduced the term quality factor. The quality factor of DRA is dependent upon the aspect ratio of DRA (rectangular DRA quality factor is dependent upon a, b, d dimensions and dielectric constant). The DRA can be represented by a parallel RLC circuit. The echoes formed inside DRA can generate higher order harmonics of fundamental frequency of input excitation. These echoes are called higher order modes. At any instant, these are a weighted sum or superposition of all the modes present inside the DRA. These are also termed dominant modes. Fundamental modes have the lowest resonant frequency, and higher order modes have higher frequency and less bandwidth with high directivity. A metal strip can be used in practical applications on DRA to play with fundamental and higher order modes. The RF signal in DRA can leak to space through transparent walls also called PMC walls. The transcendental equation is developed based on the continuity of RF signals inside and outside DRA. Also, the total electrical and magnetic energy account of DRA remains fixed. The reflection coefficient and radiation pattern parameters are measured as an outcome an antenna. The orientation of outcome fields or propagated fields are known as polarization. The circular polarization in RF signal produced by DRA can add to robustness, thus safeguarding propagated signals from certain losses. The input signal from source to microstrip line acts as a transformer. Again, signals traveling from slot to DRA act as transformers. The slot and DRA can be realized by independent equivalent RLC circuits. Each shall respond to certain resonant frequency. The microstrip line shall respond to lumped characteristics. The DRA is capable of generating resonant modes based on superposition of fields at any instant.

There can be harmonics of resonant modes because of echoes inside the DRA. These echoes generated can be called higher order modes. Now fundamental modes and higher order modes shall have to bear with a term called coupling. This coupling shall have an impact on quality factor of an antenna. The more it echoes, the more will be the value of the quality factor. The resonant frequency will shift upward for higher order modes. The resonant mode excited is a field pattern developed inside the DRA. This field pattern is fully dependent upon boundary conditions of DRA along with input excitation. Mathematically Taylor's series can provide an explanation of resonant modes and harmonics. The radio waves reflect with PEC (electrical conducting) walls and escape through PMC (magnetic conducting) walls treated as transparent walls. These radio wave propagations are based on Maxwell's equation. Poynting vector theorem can provide details of energy flow in vectored form. Their propagation obeys Maxwell's equation. Space has an impedance of 377 ohms. Horn-shaped antenna are tapered and can transform easily from 50 ohms to 377 ohms with minimal loss of RF energy. The RF fields generated can be further classified as near fields and far fields at the same frequency. The near fields have limited distance and far fields have long distance. Both these fields are used for different types of communications such Bluetooth and Wi-Fi communications. DRAs are operated on the principle of displacement current introduced in low loss high permittivity materials (8 to 125), thus reduced energy dissipation in DRA. DRA has advantage of high efficiency, even at higher operating frequency, in contrast to metal antennas. Dimensions of DRA are based on half wavelength, and TMM (thermoset microwave material) is a suitable dielectric material used for DRA. High radiation loss with low quality factor of DRA is the basis of coupling of RF signals to free space.

The terahertz frequency of smaller wavelength has advantages. Low loss and high permittivity dielectric material availability at terahertz frequencies resulted in the origin of nano DRAs. Gaussian beam input creates nonlinear radiation. Dielectric behaves like a conductor at high frequency, and a conductor behaves like dielectric at high frequency. Graphene substrate has fascinating electrical and optical nonlinear properties and possesses high plasmon sensitivity at THz frequency. Graphene with SiO_2 substrate is used in terahertz antennas. It is a THG (third harmonic generation) material. Here, third order harmonics gets generated due to Gaussian input beams. They are proximity coupled and the feed used is silver nano waveguide. Proximity feed has been used to excite higher order modes in nano DRA. The input to silver nano waveguide is fed through laser as a Gaussian beam to create SPPs (surface plasmon polaritons) into SiO_2 substrate. SPPs are optical frequencies. This phenomenon is different from e.m. waves that travel along a metal–dielectric interface. In nano DRA, the phenomenon of annihilation and creation of photons take place. Quantum electromagnetic fields are produced by current density. The second quantized current density can be built out of the Dirac field of electrons and positrons. The free electromagnetic or photon field is built out of solutions to the wave equation with coefficients being operators, namely the creation and annihilation operators of the photons. The Dirac field also consists of free waves; coefficients are operator fields in momentum space, known as the creation and annihilation fields of electrons and positrons. After computing the electromagnetic field produced by a second quantized current, the problem of determining quantum fluctuations in these fields in a given state of the electrons and positrons is deduced. The quantum electromagnetic field in nano DRA is produced by a quantum current density. The desired pattern of fluctuations or more generally, a desired pattern of higher order moments in space-time is obtained. Here, control input applied to the antenna will be in the form of a classical voltage/current source that will affect the Hamiltonian of the system.

In nano DRA, a laser is used to excite SPPs through silver nano waveguides. *Graphene* has strong light matter interaction; also it is a *third* order nonlinear response. The radiation at terahertz (THz = 10^{12} Hz) falls in between the microwave band and infrared band, i.e., 100 GHz to 10 THz. Nano DRA is useful for imaging as well as wide band and short range communications. It has large bandwidth and is useful for 5G applications. Terahertz (THz) waves are considered safe and contact free technology for biological tissue pattern study and disease detection.

This book consists of fourteen chapters and seven annexure. The book describes the design of DRA with mathematical, simulation, and experimental backgrounds. Application-oriented case studies on vehicular antennas have also been presented in this book. The book has been written in simple and lucid form.

Rajveer S. Yaduvanshi, PhD
Gaurav Varshney, PhD

Authors

Rajveer S. Yaduvanshi is currently working as a professor in the department of electronics and communication engineering in AIACTR, Government of NCT of Delhi. He was appointed as a full-time professor through UPSC examination. He earned his AMIE, ME (Hons.) from MNIT, Allahabad and PhD in ECE from Delhi University. He is also a proud fellow member of IETE. Rajveer has more than 33 years of work expertise in the teaching and design of microwave and communication engineering subjects. He has written 3 books on Antennas and published more than 100 research papers in Journals and Proceedings.

He has rendered more than one hundred invited talks on the concept of dielectric resonator antennas at both national and international levels. He has supervised 5 PhD theses, 37 MTech theses and has multiple thesis projects in his pipeline. He was also the organizer of the international conference ICPR-2017.

Prof. Rajveer is a proud recipient of academic excellence award and management excellence award in 2017. He was also selected as a director for EQDC in Gujarat Government. His research interest involves nano-dielectric resonator antennas design, RF sensors, microwave devices, millimeter wave, vehicular antennas, optical antennas, and retinal quantum antennas.

Being an avid writer and born speaker, Prof. Rajveer is a reviewer of IEEE transactions, IEEE letters, IEEE access, WPC Springer, IET and IJUWBCS journals. He is also an editorial member of MAPAN, Springer Journal.

Dr. Gaurav Varshney earned his PhD from NIT Delhi. His research interests are microwave and millimeter wave antenna designs. He has published many research papers in reputed journals and conferences.

1 Fundamentals of an Antenna

1.1 INTRODUCTION

Wireless communication has eased human life and eradicated jumbles of wires used in interconnections of electronic subsystems. Wireless sensors have made human life comfortable. Self-driving cars will be the next generation of vehicles with the invention of wireless sensors. Smart vehicles are equipped with facilities of all required infotainment and entertainment features. Communication wirelessly with self, infrastructure and surrounding vehicles requires compact, multiband and efficient antennas. Here, MIMO (multi in multi out)-based antennas can be highly directive and can have diversity [1–5]. The growth of wireless communication has created additional demand on advancement in antenna technology. An antenna is the eyes and ears to wireless communication systems. Antennas act as sensors. These are used in biomedical applications for wireless monitoring of health, radars, satellites, GPS, GNSS and even in mobile sets. The current requirements of antennas are wideband, multiband, low cost, compact, lightweight, frequency agility and aesthetic design in the era of 5G technology. Based on the requirements, one may need conformal antenna, super directive antenna, MIMO antenna or reconfigurable antenna. Hence, more and more innovations into antenna designs are required to meet the challenge. This book deals with a novel and aesthetic design of antennas for use in 5G applications [6–13].

An antenna can be defined as *a transition medium between a guided wave and free space and vice-versa*. It is a kind of transducer that converts electrical energy into electromagnetic energy and vice-versa. *Marconi* gives the idea that signals can be transmitted in free space using a radiating device called an antenna. The frequency spectrum used for different communication is given in Figure 1.1.

1.2 ANTENNA PARAMETERS

1. Frequency
2. Impedance
3. Power
4. Gain
5. Reflection coefficient (s_{11})
6. Scattering parameters
7. VSWR
8. Bandwidth

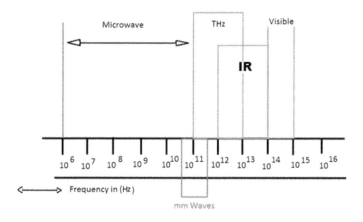

FIGURE 1.1 Frequency spectrum used for communication.

9. Beam width
10. Directivity
11. Polarization
12. Permittivity
13. Permeability
14. Dielectric constant
15. Efficiency
16. Quality factor
17. Resonance modes
18. Axial ratio
19. Aspect ratio
20. Frequency ratio
21. Low profile
22. Directive gain
23. Dipole moment
24. Higher order modes
25. Poynting vector
26. Magnetic vector potential
27. Wave number
28. Boundary conditions
29. Bessel function
30. Green function
31. Frequency spectrum
32. Conductivity
33. $\lambda = \frac{c}{f}$
34. Miniaturization
35. Low profile
36. Group delay

37. Loss tangent, $\text{Tan}\left(\partial_o\right)$
38. Free space impedance $\left(\eta = \sqrt{\frac{\mu_o}{\epsilon_o}} = 377\Omega\right)$
39. Permittivity $= 8.854 \times 10^{-12}$ F/m
40. Permeability $= 4\pi \times 10^{-7}$ H/m
41. Beam width
42. End fire radiation pattern
43. Broadside radiation pattern
44. Met material (negative permittivity)
45. Power gain
46. Multiband
47. Correlation
48. Circular polarization
49. RHCP and LHCP (right-hand circular polarization and left-hand circular polarization)
50. MIMO (multi input multi output)
51. Envelopment correlation coefficient (ECC)
52. Directive gain (DG)
53. Mean effective gain (MEG)
54. Channel capacity loss (CCL)
55. Total active reflection coefficient (TARC).
56. Radiated
57. Polarization (linear, elliptical, circular, LHCP, RHCP)
58. Frequency ratio
59. Low profile
60. Dipole moments
61. Group delay
62. Loss tangent
63. End fire radiation pattern
64. Broadside radiation pattern
65. Bore sight radiation pattern

1.3   RADIATION MECHANISM OF THE ANTENNA

Since the invention of the antenna, a number of antenna structures have been designed and implemented. The different antennas have been designed for different applications. The radiation mechanism of the antenna can be explained using the basic radiation equation given as:

$$\frac{di}{dt} \times l = q_v \times \frac{dv_z}{dt} \tag{1.1}$$

For the radiation from an antenna, it is necessary to have the time varying current and oscillating charge. The time varying current shows the steady state harmonic motion and charge shows the transients or pulses. The radiation is perpendicular to the acceleration. The radiated power is equal to the square of any single side of equation (1.1).

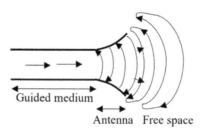

FIGURE 1.2   Radiation mechanism of the antenna.

Consider a two-wire transmission line with a distance smaller than the operating wavelength between the wires shown in Figure 1.2. The Figure 1.2 represents the radiation mechanism of the antenna. We then consider the lossless transmission line having resistance R, inductance L, conductance G and capacitance C. If there is a generator connected at the source end of the transmission line, the energy is guided from the source end to the load end. This energy consists of two parts: electric and magnetic. The electric energy $(\frac{1}{2}CV^2)$ will be stored by the capacitor and magnetic energy $(\frac{1}{2}LI^2)$ by the inductor.

Consider the open end after the transition medium of the transmission line. The current at the load end will be zero due to the open-ended transmission line. The total energy must be conserved because it can neither be generated nor destroyed. Hence, the total energy stored in the capacitor increases due to rise in the voltage at the open end of the transmission line. The increase in the voltage at the open end of the transmission line generates a current, which flows from the load end to the source end of the transmission line. This is considered as the reflected current. The same happens in the case of the short-circuited transmission line in which a rise in current generates the voltage waves in the reverse direction.

In the case of the open-ended transmission line, the current momentarily remains zero. This can be considered as the time taken by incident current waves to revert. During this small time period, some of the energy is leaked out in the space in form of radiation. The radiation mechanism is displacement current due to excitation circulating into the RDRA or dielectric medium. This dielectric material permittivity ϵ_r forms capacitance, then $= \frac{\varsigma}{f}$, hence charge into certain arrangements can define energy stored into RDRA. This generates a field pattern into the bounded region, thus creating a magnetic field inside resonator. A physical phenomenon seems to happen on if a metal wall is penetrating through dielectric medium; expansion and contraction of dielectric medium will take place, resulting into acceleration of medium particles. This creates a fringing effect at the side wall and energy radiates into the medium. This partially transparent magnetic wall allows fields to escape. This phenomenon is known as Cherenkov radiation.

$n \times H = 0$	H Tangential	PMC
$n \cdot E = 0$	E Normal	
$n \times E = 0$	E Tangential	PEC
$n \cdot H = 0$	H Normal	

Emission of three radiations by oscillation currents subjected to RDRA as excitation can be used for analysis of time dependence. Each component can be separately obtained by separation of variables for various studies. No generality is lost by the system in terms of potential, fields and radiation from system of charges and currents. Varying sinusoidal in time energy flow in a particular direction can be treated as power radiated per unit solid angle (energy per unit area per unit time):

$$\left.\begin{array}{l} H = n \times E \\ E = H \times n \end{array}\right\} \text{ by duality}$$

Relative amplitude of wave can be expressed by $d(mnp)$ and $c(mnp)$, the solution λ, k, ψ the wavelength, wave meter and wave functions define eigen value based on boundary specified.

$$d = \sum e \cdot r$$

where:
  d is the dipole moment
  e is the charge
  r is the distance between two charges

$$d = \frac{d}{dt} \sum e \cdot r = \sum e \cdot v$$

$$d^{\cdot} = \frac{d}{dt} \sum e \cdot v$$

Hence, charges can radiate if they move with acceleration and there will be no radiation even if they move with fixed/uniform velocity/motion.

Hence, pointing vector gives the following energy flow:

$$S = c \cdot \frac{H^2}{4\pi} \cdot n$$

1.3.1 RADIATION RESISTANCE

The radiation resistance is a virtual resistance which couples the energy from the antenna to the free space (Kraus, 2010). In the case of the receiving antenna, it receives the active as well as passive radiation. The rise in the temperature of the objects from which radiation is coming increases the temperature of the receiving

FIGURE 1.3   The radiation resistance of the antenna.

antenna. Consequently, there is an increment in the radiation resistance. Thus, an antenna can be used as the temperature-sensing device. Figure 1.3 shows the radiation resistance.

1.4 PARAMETERS OF AN ANTENNA

The basic antenna parameters are bandwidth, gain, radiation pattern and radiation efficiency. These parameters are described as follows:

1.4.1 BANDWIDTH

The bandwidth of the antenna can be defined in terms of impedance matching, gain and polarization. The bandwidth can be defined as the difference between the higher and lower cut off frequency for an acceptable limit of the parameter.

1.4.2 REFLECTION COEFFICIENT

The reflection coefficient is the complex quantity having magnitude and phase. The reflection coefficient of the transmission line can be defined as the ratio of complex amplitude of the reflected wave to the complex amplitude of incident wave.

$$\Gamma = \frac{E_r}{E_i} \tag{1.2}$$

Suppose we have a transmission line having a generator connected at the source end and load impedance Z_l connected at the load end. The reflection coefficient of the antenna is given by

$$\Gamma = \frac{Z_l - Z_o}{Z_l + Z_o} \tag{1.3}$$

where, Z_o is the characteristic impedance of the transmission line. The impedance mismatch creates the reflected waves. The reflection coefficient varies from 0 to 1.

1.4.3 RETURN LOSS

The ratio of the reflected power (P_r) to the incident power (P_i) is defined as the return loss (RL). To calculate the return loss, we consider the average power. This is given by

$$RL = |\Gamma|^2 = \frac{\langle P_r \rangle}{\langle P_i \rangle} \tag{1.4}$$

$$RL[dB] = 10\log\left(|\Gamma|^2\right)$$

$$RL[dB] = 20\log|\Gamma| \tag{1.5}$$

One must be clear about the reflection coefficient and return loss that the reflection coefficient is always negative in dB and return loss is always positive. Figure 1.4a shows the reflection coefficient and Figure 1.4b shows the return loss of an antenna.

Figure 1.4a shows the impedance bandwidth (BW). The impedance BW can be defined as the operating frequency band between the upper (f_h) and lower (f_l) cut-off frequencies in the plot of reflection coefficient $(S_{11} \leq -10\,\text{dB})$. The value of the reflected power can be calculated as follows:

$$-10 = 10\log\left(\frac{P_r}{P_i}\right)$$

$$P_r = 0.1 \times P_i$$

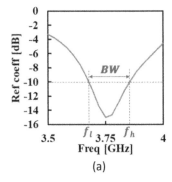
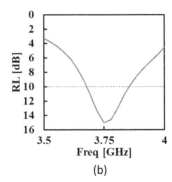

FIGURE 1.4 Frequency response of an antenna (a) reflection coefficient and (b) return loss.

For the value of reflection coefficient −10dB, the reflected power is 10% of the incident power. In other words, 90% of the power is radiated in free space. Any system is considered to be in the steady state if its response is more than 90% or has the 10% tolerance band for the power loss.

The impedance bandwidth of an antenna can be defined as the frequency band where the reflection coefficient remains less than or equal to −10dB. The impedance bandwidth is shown in Figure 1.4a. It can be defined as the fractional bandwidth as given by

$$\text{Impedance bandwidth}\left(BW_{lm}\right) = \frac{f_h - f_l}{\left(f_h + f_l\right)/2} \times 100\% \qquad (1.6)$$

Since the reflection coefficient is dependent on the load and characteristic impedance, the bandwidth is called the impedance bandwidth.

1.5 RADIATION PATTERN

Radiation pattern is the representation of the field or power in space as a function of elevation angle (θ), azimuth angle (ϕ), and distance (r). The radiation pattern has three components as given by

1. E_θ,
2. E_ϕ, and
3. The phases of the field components.

All these components vary with θ and ϕ.

To understand the radiation pattern, it is necessary to go through the coordinate system representing a point P in free space. Figure 1.5a shows the resolution of the position vector $\vec{r}$ in a spherical coordinate system. Figure 1.5b shows the formation of a cylinder when the constant ϕ-plane rotates around the z-axis from 0 to 2π. Figure 1.5c shows the formation of the cone by the rotation of position vector $\vec{r}$ around z-axis.

Figure 1.6 shows the 3D radiation pattern of an antenna. Cutting the 3D radiation pattern in xz and yz-planes provides the 2D radiation pattern as shown in Figure 1.6a and b, respectively.

Combining these two plots shown in Figure 1.7a and b, we can find two wireframes of the xz and yz-plane radiation patterns on a single 3D plot. Rotating this combined plot around the z-axis, we can find the full 3D radiation pattern.

The analysis of radiation pattern defines some points like cross polarization, front-to back ratio and beam width. *The cross polarization can be defined as the unwanted field component in the radiated field*. In the case of a linearly polarized (LP) antenna radiating with vertical polarization, the presence of the horizontal component introduces the cross polarized component of the radiated field. Similarly, in the case of a circularly polarized (CP) antenna, if the antenna is designed to provide the right-hand CP (RHCP) field, the left-hand CP (LHCP) field component acts as

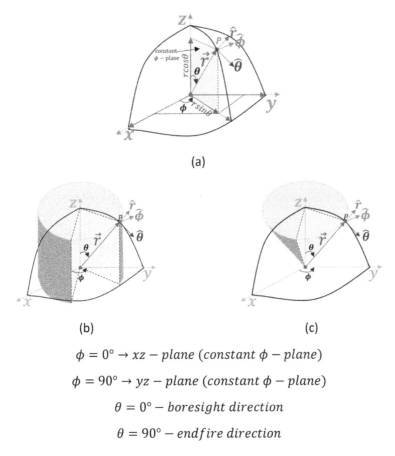

(a)

(b)　　　　　　　　　　　　　　　(c)

$$\phi = 0° \rightarrow xz - plane \ (constant \ \phi - plane)$$

$$\phi = 90° \rightarrow yz - plane \ (constant \ \phi - plane)$$

$$\theta = 0° - boresight \ direction$$

$$\theta = 90° - endfire \ direction$$

FIGURE 1.5 (a) Spherical coordinate system, (b) cylinder formation by the rotation of constant ϕ-plane, and (c) cone formation by the rotation of point P around z-axis.

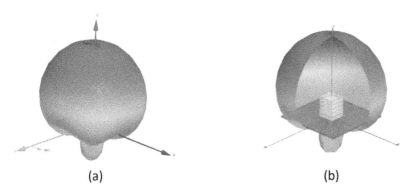

(a)　　　　　　　　　　　　　　　(b)

FIGURE 1.6 (a) 3D radiation pattern of an antenna and (b) cross sectional view of the radiation pattern.

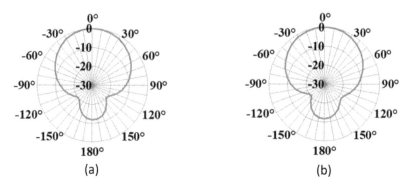

FIGURE 1.7 Total radiated E-field pattern in free space in (a) xz-plane and (b) yz-plane.

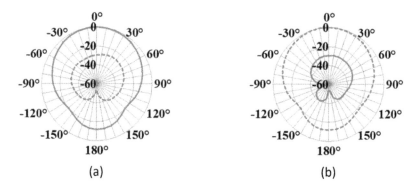

FIGURE 1.8 Radiation pattern of the antenna in (a) xz-plane and (b) yz-plane (solid line: E_θ component; dotted line: E_ϕ component).

the cross polarized component. The separation between co- and cross-polarized field components more than 20 dB shows the good radiation characteristics of the antenna. In Figure 1.8, the E_θ and E_ϕ components of the fields are shown. It can be observed that the E_θ component is dominant in xz-plane while the E_ϕ component is dominant in yz-plane. This confirms the LP radiations as well as a good radiator since the cross polarization separation in both principal planes is more than 20dB.

1.6   BEAM WIDTH OF AN ANTENNA

Theta angle $(\theta) = [(\lambda)$ Wavelength$/(D)$ Diameter of antenna$]$ in radians

The beam width can be categorized as the half-power beam width (HPBW) and full-null beam width (FNBW). This is the angular width between the half-power points from the maxima in the case of HPBW and between the two nulls of the power in case of FNBW. Figure 1.9 shows the normalized radiation pattern in xz and yz-planes.

$$E_\theta\left(\theta,\phi\right)_n = \frac{E_\theta(\theta,\phi)}{E_\theta(\theta,\phi)_{\max}} \tag{1.7}$$

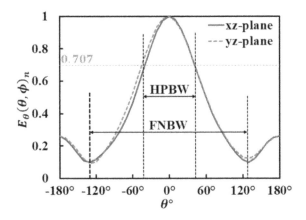

FIGURE 1.9 Normalized radiation pattern.

The HPBW is 2θ and $E_\theta(\theta,\phi)_n = 0.707$ at HPBW. The value of $E_\theta(\theta,\phi)_n$ is 0 for FNBW.

The power pattern can be drawn in terms of the power per unit area. The normalized power pattern can also be derived using field pattern as given by

$$P_n(\theta,\phi) = \frac{S(\theta,\phi)}{S(\theta,\phi)_{max}} \tag{1.8}$$

where $S(\theta,\phi)$ is the Poynting vector and $S(\theta,\phi) = \dfrac{\left[E_\theta^2(\theta,\phi) + E_\phi^2(\theta,\phi)\right]}{Z_o}$ W/m^2 and Z_o is the free space intrinsic impedance.

Suppose an antenna has the field pattern $E(\theta) = \sin^2\theta$; $0 \le \theta \le 90°$; the HPBW will be calculated as $E(\theta) = 0.707 = \cos\theta \to \theta = 45°$. Hence, the HPBW is $2\theta = 90°$. Similarly, we can calculate the value of the FNBW as $E(\theta) = 0 = \cos\theta \to \theta = 90°$. Hence, the FNBW is $2\theta = 180°$. Furthermore, we can find the shape of the radiation pattern if FNBW and HPBW is already known.

1.6.1 3D Beam Width and Beam Area

A solid beam can be said to be a 3D beam width.

The beam area of the radiation pattern can be calculated as shown in Figure 1.10. The beam area is given as follows:

$$dA = r^2\sin\theta d\theta d\phi \tag{1.9}$$

$$dA = r^2 d\Omega$$

$$d\Omega = \sin\theta d\theta d\phi$$

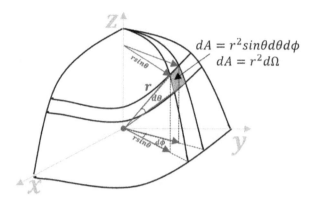

FIGURE 1.10   Spherical coordinate system showing the beam area.

where Ω is the solid angle subtended by area dA. The solid angle subtended by the sphere is 4π. Hence, the area of a sphere is $4\pi r^2$. The unit of the solid angle is steradian or square degree.

$$1 \text{ Steradian} = (\text{degree})^2 = \left(\frac{180}{\pi}\right)^2 = 3282.8064 \text{ square degrees}$$

$$4\pi \text{ steradians} = 41252.96 \text{ square degrees} \approx 41253 \, \Omega$$

The beam area or beam solid angle is the integral of the normalized power pattern over a sphere having solid angle $4\pi \, sr$, and given as:

$$\Omega_A = \int_\theta \int_\phi P_n(\theta,\phi)\sin\theta \, d\theta \, d\phi = \iint_{4\pi} P_n(\theta,\phi) \, d\Omega(sr) \qquad (1.10)$$

Ω_A is the solid angle through which power $P(\theta,\phi)$ is radiated by the antenna and remains maximum and zero otherwise. Thus, the power radiated is $P(\theta,\phi)\Omega_A$.

The beam area can be approximated by neglecting the minor lobes if the half-power beam widths of the principal plane are already known.

$$\Omega_A = \theta_{\text{HPBW}}\phi_{\text{HPBW}} \qquad (1.11)$$

1.6.2   RADIATION INTENSITY

This is the power radiated by the antenna per unit solid angle. The unit of the radiation intensity is Watts/sr. This is independent of the distance unlike the pointing vector (both are assumed in the far-field). The normalized power pattern can also be calculated using the radiation intensity.

$$P_n(\theta,\phi) = \frac{U(\theta,\phi)}{U(\theta,\phi)_{max}} = \frac{S(\theta,\phi)}{S(\theta,\phi)_{max}} \tag{1.12}$$

1.6.3 BEAM EFFICIENCY

The total beam area (Ω_A) is the summation of the minor lobe area (Ω_m) and major lobe area (Ω_M). Thus, the beam efficiency is given as follows:

$$\text{Beam efficiency}\left(\epsilon_M\right) = \frac{\Omega_M}{\Omega_A} \tag{1.13}$$

where $\Omega_A = \Omega_m + \Omega_M$.

1.6.4 STRAY FACTOR

The ratio of the minor to total beam area is known as the stray factor. This factor can be expressed mathematically as follows:

$$\text{Stray factor}\left(\epsilon_m\right) = \frac{\Omega_m}{\Omega_A} \tag{1.14}$$

$$\epsilon_M + \epsilon_m = 1 \tag{1.15}$$

1.6.5 DIRECTIVITY

The directivity is defined as the ratio of the maximum power density to the average power density over a sphere. This is dimensionless and always greater than or equal to unity.

$$D = \frac{P(\theta,\phi)_{max}}{P(\theta,\phi)_{av}} \tag{1.16}$$

$$P(\theta,\phi)_{av} = \frac{1}{4\pi} \int_0^{2\pi} \int_0^{\pi} P(\theta,\phi)\sin\theta d\theta d\phi \tag{1.17}$$

$$P(\theta,\phi)_{av} = \frac{1}{4\pi} \int_0^{2\pi} \int_0^{\pi} P(\theta,\phi) d\Omega \text{ W/sr} \tag{1.18}$$

$$D = \frac{P(\theta,\phi)_{\text{max}}}{\dfrac{1}{4\pi} \displaystyle\int_0^{2\pi} \int_0^{\pi} P(\theta,\phi)d\Omega} \tag{1.19}$$

$$D = \frac{1}{\dfrac{\dfrac{1}{4\pi} \displaystyle\int_0^{2\pi} \int_0^{\pi} P(\theta,\phi)d\Omega}{P(\theta,\phi)_{\text{max}}}} \tag{1.20}$$

$$D = \frac{4\pi}{\Omega_A} \tag{1.21}$$

$$D = \frac{4\pi}{\theta_{\text{HPBW}}\phi_{\text{HPBW}}} \tag{1.22}$$

$$D = \frac{41253}{\theta_{\text{HPBW}}\phi_{\text{HPBW}}} \tag{1.23}$$

Thus, directivity can also be calculated by the beam area and beam widths in the principal planes.

For an isotropic antenna, $\Omega_A = 4\pi \ (sr) \to D = 1$.

For a short dipole, $\Omega_A = 2.67\pi \ (sr) \to D = 1.5$ (1.76dBi).

Thus, minimum directivity is unity.

Equation 1.23 shows that the directivity and beam width product of an antenna is always constant. Wider beam width leads to the smaller directivity and vice-versa.

1.7 GAIN

The antenna gain, G, is always less than directivity due to losses (ohmic) associated with the antenna. It can be measured in terms of the reference antenna if gain is known. It is given by:

$$G \text{ (Test antenna)} = \frac{P_{\text{max}}(\text{Test antenna})}{P_{\text{max}}(\text{Reference antenna})} \times G \text{ (Reference antenna)} \tag{1.24}$$

where P_{max} is the maximum power density. Generally, a reference antenna is an isotropic antenna which is hypothetical only. Hence, a short dipole is used as a reference antenna. The gain of the antenna is a unitless quantity. Generally, the gain is measured in dB or dBi (decibels over isotropic).

1.8   ANTENNA RESOLUTION

The resolution of the antenna is defined as the capability to resolve the number of signals from different sources. Generally, the resolution of the antenna is equal to half of full-null beamwidth.

$$\frac{\text{FNBW}}{2} \approx \text{HPBW} \tag{1.25}$$

These, N number of signals received in phase from different sources will enhance directivity D. This is given by the following formulation:

$$N = D = \frac{4\pi}{\Omega_A} \tag{1.26}$$

It is important to note that the left side of equation (1.25) is little greater than the right side term and gives more accurate measure to the beam area.

The polarization of waves is the arrangement of the field vectors in space. It can be categorized as linear, circular and elliptical. The most general case of polarization is elliptical polarization (Figure 1.11).

1.8.1   LINEAR POLARIZATION

In case of the linear polarization, the field vectors are arranged in a single direction, i.e., y-axis and plane wave is propagating in the outward direction along the positive z-axis. The plane wave can be represented as follows:

$$E = E_1 \sin(\omega t - \beta z) \tag{1.27}$$

where β is the propagation constant and E_1 is the amplitude of the wave.

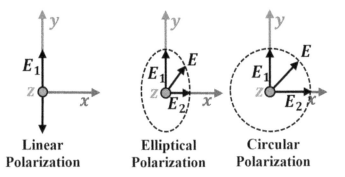

Linear Polarization   **Elliptical Polarization**   **Circular Polarization**

FIGURE 1.11   Representation of the different types of polarization.

In case of the elliptical polarization, there are two components of the field in x and y direction. The phase difference between them is δ. Both components of the field are having different amplitude $(E_1 \neq E_2)$. In this case, the resultant field vector rotates as a function of time. If the amplitude of both the field components is equal then the wave is considered circularly polarized.

The elliptically or circularly polarized waves can be represented as the combination of two linearly polarized waves.

$$E_x = E_1\sin(\omega t - \beta z) \tag{1.28}$$

$$E_y = E_2\sin(\omega t - \beta z + \delta) \tag{1.29}$$

Here, E_1 and E_2 are the amplitude of the linearly polarized components in x and y direction, respectively (Figure 1.12). Combining equations (1.28) and (1.29), the resultant field vector is given as follows:

$$E = \hat{x}E_1 \sin(\omega t - \beta z) + \hat{y}E_2\sin(\omega t - \beta z + \delta) \tag{1.30}$$

$$E = \hat{x}E_1 \sin(\omega t) + \hat{y}E_2 \sin(\omega t + \delta)|_{z=0} \tag{1.31}$$

At $z = 0$, $E_x = E_1 sin\omega t$ and $E_y = E_2 \sin(\omega t + \delta) = E_2(\sin\omega t \cos\delta + \cos\omega t \sin\delta)$. Thus, solving these equations we get $\sin\omega t = \frac{E_x}{E_1}$ and $\cos\omega t = \sqrt{1 - (E_x / E_1)^2}$.

Now solving these equations, we can get the equation representing the ellipse as given by

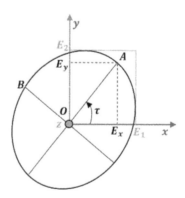

FIGURE 1.12 Polarization ellipse showing x and y components of the field with the peak values E_1 and E_2, respectively and tilt angle τ.

$$E_y = E_2 \frac{E_x}{E_1} \cos\delta + E_2 \left(\sqrt{1 - \left(\frac{E_x}{E_1}\right)^2} \right) \sin\delta$$

$$\left(\frac{E_y}{E_2} - \frac{E_x}{E_1} \cos\delta\right)^2 = \left(1 - \left(\frac{E_x}{E_1}\right)^2\right) \sin^2\delta$$

$$\frac{E_y^2}{E_2^2} + \frac{E_x^2}{E_1^2} \cos^2\delta - \frac{2E_x E_y \cos\delta}{E_1 E_2} = \left(1 - \left(\frac{E_x}{E_1}\right)^2\right) \sin^2\delta$$

$$\frac{E_y^2}{E_2^2} + \frac{E_x^2}{E_1^2} - \frac{2E_x E_y \cos\delta}{E_1 E_2} = \sin^2\delta \tag{1.32}$$

In equation (1.32), all are constants except E_x and E_y. This equation represents the polarization ellipse. τ is the tilt angle of the ellipse.

1. For $E_1 = E_2$, $\delta = 0°$, linearly polarized wave with resultant arranged at angle 45°.
2. For $E_1 = 0$; LP wave along y-axis.
3. For $E_2 = 0$; LP wave along x-axis.
4. For $E_1 = E_2$; $\delta = +90°$; LHCP wave.
5. For $E_1 = E_2$; $\delta = -90°$; RHCP wave.

For a plane wave, the phase of the wave front remains constant, i.e., $\omega t - \beta z = \phi = $ constant. The plane wave is the function of z and t. This gives the relation for the function of time;

$$\omega = \frac{\beta dz}{dt} = \beta v_p \tag{1.33}$$

where v_p is phase velocity of the plane wave front in $+z$ direction. The phase velocity of the plane wave is given as $v_p = 1/\sqrt{\mu\epsilon}$. For any medium $\mu = \mu_o$ and $\epsilon = \epsilon_r \epsilon_o$, gives the relation $v_p = \frac{c}{\sqrt{\epsilon_r}} = \frac{c}{n}$, where, ϵ_o is permittivity, μ_o is permeability of free space, ϵ is permittivity of the medium, ϵ_r and n are relative permittivity and refractive index of the medium, respectively.

If the phase is the function of position in space, then

$$\frac{d\omega}{dz} = \beta \qquad (1.34)$$

The Poynting vector of a circularly and elliptically polarized wave can be calculated as:

$$S = \frac{1}{2}\left(E \times H^*\right) \qquad (1.35)$$

The average Poynting vector is the real part of equation (1.36):

$$S_{av} = \frac{1}{2}Re\left(E \times H^*\right) = \frac{\hat{z}\left(E_1^2 + E_2^2\right)}{2Z_o} = \frac{\hat{z}E^2}{2Z_o} \qquad (1.36)$$

$E_1^2 + E_2^2 = E^2$; E is the resultant field amplitude.

1.9   AXIAL RATIO

The axial ratio (AR) is defined as the ratio of major axis to minor axis of an ellipse. From Figure 1.11, it is given by:

$$AR = \frac{OA}{OB} \qquad (1.37)$$

AR = 1; circular polarization
AR = ∞; linear polarization

In general, it is difficult to achieve AR = 1. Hence, in the case $OB = \frac{OA}{\sqrt{2}} \rightarrow AR = 3dB$, we consider the criteria for obtaining the CP waves from an antenna.

1.9.1   AR BANDWIDTH

The frequency band over which AR remains less than or equal to 3dB is called the AR bandwidth. Any antenna is considered circularly polarized if both impedance passband ($S_{11} \leq -10dB$) and AR passband (AR $\leq$ 3dB) overlap each other.

1.10 MODERN CLASS OF ANTENNAS

Various modern types of antennas based on different structures, mechanisms and applications are discussed in the following section. In the current wireless communication systems, the wire, microstrip patch and DRA are being implemented and utilized.

1.10.1 WIRE ANTENNAS

Wire antennas are very common and used in many applications such as automobiles, ships, aircrafts, spacecraft and others. Wire antennas are of many configurations such as in the form of dipole, loop and helix. The dipole antenna is a type of wire antenna as shown in Figure 1.13 (Balanis, 1997) (Figures 1.14 through 1.16).

Antennas are used as transmistter antenna as well as receiver antenna.

FIGURE 1.13   Dipole antenna.

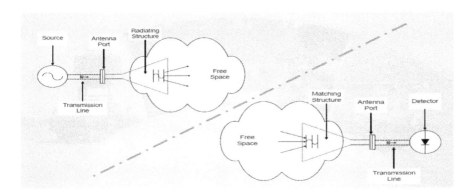

FIGURE 1.14   Antenna in trans/receive mode.

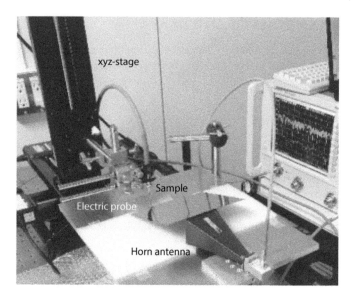

FIGURE 1.15   Antenna used as test antenna.

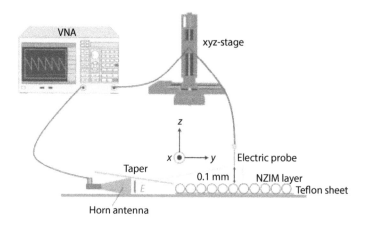

FIGURE 1.16   VNA used in testing antenna parameters.

1.11   MEASUREMENTS OF ANTENNA

1.11.1   Scattering Parameters of a Two-Port Microwave Network

To understand the electrical behavior of a microwave network, scattering parameters (S-parameters) are used (Kurokawa, 1965). The S-parameters are similar to the other network parameters like z, h, y and $ABCD$-parameters. The main distinguishing factor of S-parameters from other parameters is that they use matched load condition instead of the open and short-circuit condition. The S-parameters are used for networks operating at higher frequencies as open and short circuit leads to the mismatched impedance that causes high-reflected powers. Hence, S-parameters are used for the antenna measurements in power terms. Various antenna parameters like return loss (RL), reflection coefficient, and voltage standing wave ratio (VSWR) have been calculated using S-parameters. When an electromagnetic wave travels across the antenna through a transmission line, several discontinuities in the network come into effect. The introduction of antenna within the transmission line network causes a change in the wave impedance and characteristic impedance. These discontinuities from the transmission line to the antenna can be easily measured and compensated using the S-parameter. Thus, S-parameters are mainly used in microwave networks where current and voltage measurements are affected greatly due to the insertion of other elements. In this case, energy or signal power constraints are easily calculated using S-parameters. S-parameters are calculated in the matrix form depending upon the number of ports used in the microwave networks. The number of ports in the electrical network defines the points at which the electrical signal is incoming or outgoing in the network.

1.11.2 S-PARAMETER MATRIX FOR A MICROWAVE NETWORK

For N-port network, let us consider a port i chosen from 1 to N number of ports where a_i and b_i are the incident and reflected power of the ports, respectively. The relation of these incident and reflected powers can be defined in terms of voltage, current and impedance of the network (Kurokawa, 1965).

Incident power at the i_{th} port is

$$a_i = \frac{1}{2} k_i \left(V_i + Z_i I_i \right) \tag{1.38}$$

Reflected power at the i_{th} port is

$$b_i = \frac{1}{2} k_i \left(V_i - Z_i^* I_i \right) \tag{1.39}$$

where V_i, I_i are amplitudes of voltage and current of the i_{th} port, respectively. Z_i is the impedance at port i with complex conjugate Z_i^*. The term k_i is expressed as follows:

$$k_i = \frac{1}{\sqrt{\left| Re\{Z_i\} \right|}} \tag{1.40}$$

Let us assume that the reference impedance Z_0 is same for all the ports of a microwave network. In this condition, incident and reflected power have been simplified as follows:

$$a_i = \frac{1}{2}\frac{V_i + Z_o I_i}{\sqrt{Re\{Z_i\}}} \tag{1.41}$$

$$b_i = \frac{1}{2}\frac{V_i - Z_o^* I_i}{\sqrt{Re\{Z_i\}}} \tag{1.42}$$

The relation between incident and reflected power can be expressed using the S-parameter matrix as:

$$[b] = [S]\cdot[a] \tag{1.43}$$

1.11.3 Two-Port Microwave Network

By using the equation (1.43), the S-matrix equation for the two-port system is expressed as follows (Figure 1.17).

$$\begin{bmatrix} b_1 \\ b_2 \end{bmatrix} = \begin{bmatrix} S_{11} & S_{12} \\ S_{21} & S_{22} \end{bmatrix} \begin{bmatrix} a_1 \\ a_2 \end{bmatrix} \tag{1.44}$$

where:
  S_{11} is the input port voltage reflection coefficient,
  S_{12} is the reverse voltage gain,
  S_{21} is the forward voltage gain, and
  S_{22} is the output port voltage reflection coefficient.

On expanding the metrics, we get

$$b_1 = S_{11}a_1 + S_{12}a_2 \tag{1.45}$$

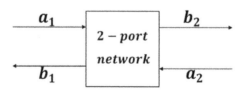

FIGURE 1.17   Two-port network.

$$b_2 = S_{21}a_1 + S_{22}a_2 \qquad (1.46)$$

Equations (1.45) and (1.46) show the relationship between incident and reflected power at port 1 and 2. If input power a_1 is given at port 1, the power either exits from port 2 (b_2) or returns back from the same port 1 (b_1) and vice versa for port 2. If port 2 is terminated at Z_0 impedance $(a_2 = 0)$, then the power b_2 at port 2 will be absorbed totally irrespective of the power a_1 as shown in equation (1.46).

1.11.4 PROPERTIES OF TWO-PORT NETWORKS

1.11.4.1 Return Loss

Return loss is the measure of the closeness between the input impedance of the network and the overall impedance of the system.

$$RL_{in} = -20\log_{10}|S_{11}|\, dB \qquad (1.47)$$

1.11.4.2 Insertion Loss

Insertion loss is the measure of the magnitude of the transmission coefficient. It is expressed in decibels.

$$IL = -20\log_{10}|S_{21}|\, dB \qquad (1.48)$$

1.11.4.3 Reflection Coefficient

S_{11} and S_{22} are the reflection coefficient parameters. The reflection coefficient at the input and output port is expressed as follows:

$$\Gamma_{in} = S_{11}, \ \Gamma_{out} = S_{22} \qquad (1.49)$$

1.11.4.4 Voltage Standing Wave Ratio

Voltage standing wave ratio (VSWR) is the ratio of the standing wave maximum voltage to the minimum voltage. It is directly related to the reflection coefficient value. VSWR is a scalar quantity and is expressed as the ratio of S_{11} and S_{22}.

$$S_{in} = \frac{1+|S_{11}|}{1-|S_{11}|} \qquad (1.50)$$

$$S_{out} = \frac{1+|S_{22}|}{1-|S_{22}|} \qquad (1.51)$$

1.12   DRA PARAMETERS: OPERATING PRINCIPLE OF DRA

There are three types of boundary conditions that can be utilized to explain the operating principle of the DRA. The three media are already explained as the electric wall, magnetic wall, and dielectric interface (Pozar, 2005).

1. In the case of an electric wall, the Maxwell's equations can be written as

$$\hat{n} \cdot D = \rho_s \tag{1.52}$$

$$\hat{n} \cdot B = 0 \tag{1.53}$$

$$\hat{n} \times E = 0 \tag{1.54}$$

$$\hat{n} \times H = J_s \tag{1.55}$$

2. In the case of a magnetic wall boundary condition, the tangential components of the magnetic field are absorbed by the magnetic surface.

$$\hat{n} \cdot D = 0 \tag{1.56}$$

$$\hat{n} \cdot B = 0 \tag{1.57}$$

$$\hat{n} \times E = -M_s \tag{1.58}$$

$$\hat{n} \times H = 0 \tag{1.59}$$

3. The dielectric interface can be considered as the coupled region where the fields are coupled from one dielectric medium to another dielectric medium.

$$\hat{n} \cdot D_1 = \hat{n} \cdot D_2 \tag{1.60}$$

$$\hat{n} \cdot B_1 = \hat{n} \cdot B_2 \tag{1.61}$$

$$\hat{n} \times E_1 = \hat{n} \times E_2 \tag{1.62}$$

$$\hat{n} \times H_1 = \hat{n} \times H_2 \tag{1.63}$$

In the case of DRA, the main radiator is made up of ceramic or insulator hence we cannot apply the electric wall boundary conditions. The another approach is to assume the magnetic wall boundary conditions around the DR which were adopted by researchers initially. The problem with the magnetic wall boundary conditions is that the tangential component of the magnetic field is the only source of energy transfer from DR to free space that becomes zero. Hence, the magnetic wall boundary conditions cannot be applied at the outer surface of the DR. The only boundary condition which can be fitted with the operation of DR is the coupled region boundary condition.

The generation of electromagnetic radiation is obtained by the oscillations of electrons, which causes acceleration and de-acceleration. Hence, electromagnetic wave comes into existence. The DRA operation is based on the phenomenon of how DR radiates when excitation is applied. Richtmyer reported that DR must radiate in free space (Richtmyer, 1939). It considers the periodic solution of Maxwell's equations of DR placed in free space. He explained the concept by considering that DR is placed in a spherical coordinate system. The origin is on or near the DR with the coordinates of the system designated as r, θ and ϕ. The resonator is of finite dimensions, and a sphere can be considered around it having radius, R. Using the concept of vector wave equations in spherical coordinate system, the field can be considered in terms of a fundamental set of solutions applied outside the sphere. For this fundamental set of solutions, the power is considered proportional to the sum of squares of their coefficients. This implies that zero power means all the coefficients of the solutions have vanished. Using the principle of analytical continuation, the field outside the resonator must vanish identically in the free space or just outside the resonator. Using the boundary condition on the outer surface of the DR implies that the tangential component of electric field intensity (E) and normal component of electric displacement (D) must be continuous. Hence, the field must vanish just outside the resonator since dielectric constant is assumed finite. This implies that the non-radiating fields are zero everywhere. The wave equation is of second order in time derivative, and it is important to be considered. But by the nature of electromagnetic waves, the field cannot vanish outside the resonator if oscillations exist. The field may lie outside the resonator by current distribution due to the motion of ions inside the DR, due to which, oscillations going outside the DR result in radiation. This phenomenon proves that DR radiates.

When DRA is excited, the radio waves are fed inside the DR and bounce back and forth within the resonating walls resulting in standing waves. These standing waves store the energy in the form of electric field components E and magnetic field components H. Due to the accelerating current, the time-varying field radiates away from the DRA in the free-space. Due to fringing effect, the magnetic field leaks out into the environment through the walls of DRA into free-space. Thus, DRA radiates.

1.12.1 Advantages of DRA

The DRA offers simple geometry, small size, high radiation efficiency, high gain, low losses, and improved bandwidth over other antennas (Mongia & Bhartia, 1994). The DR antennas have many advantages over the microstrip patch antenna such as no inherent conduction losses due to the dielectric radiator. Some of the important advantages of DRAs are depicted as follows.

1. DRAs can be easily fabricated with different geometries like rectangular, circular, hemispherical, cylindrical, etc. Also, the fabrication of such geometries can be easily done.

2. The DR dimension is of the order $\lambda_0/\sqrt{\epsilon_r}$, where λ_0 is the wavelength in free-space and ϵ_r is the dielectric constant of the DR material. Hence, the size of the antenna can be reduced by choosing the higher value of dielectric constant.

3. The modes in the DRA depend on the shape and aspect ratio of the DR. This will provide the degree of flexibility within the structure as different radiation patterns can be obtained through these modes according to the requirements.

4. DRAs offer no inherent conduction losses, as the resonator is non-metallic. Due to this advantage, DR can be used for the fabrication of millimeter wave antennas.

5. DRAs are easy to integrate with other existing technologies. This is because they offer simple coupling schemes to all the feeding techniques.

6. The dielectric strength of DR is quite high, which provides good handling capacity even at higher power. Moreover, the temperature-stable ceramics enable the antenna to operate in a wide temperature range. There is no frequency drift by change in the temperature of DRA.

7. The radiation characteristic of the antenna can be explained using the different modes excitation of the DR structure. Each mode inside the DR has a unique field pattern depending upon the external field distribution.

8. The Q-factor of the resonator reduces in the open environment, which increases the antenna bandwidth. In addition, the bandwidth can easily be varied by changing the various parameters of the DRA.

1.12.2 Important Parameters of DRA

Various parameters play an important role in the study of DRAs. These parameters include dielectric constant, Q-factor, resonant modes, etc. All these parameters are discussed in detail in the following section (Figure 1.18).

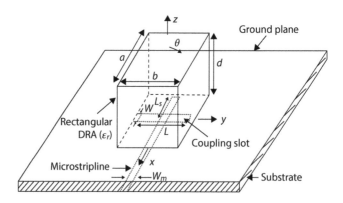

FIGURE 1.18 Rectangular DRA.

1.12.3 Nano DRA (Figures 1.19 through 1.21)

1.12.4 Dielectric Constant

Dielectric constant is defined as "the ratio of the amount of electrical energy stored inside the material when an external voltage is applied to that of the energy stored in a vacuum." It is denoted as ϵ_r. The dielectric constant of a material concentrates on the electrostatic lines of flux under the given conditions. Dielectric material supports the E-field when energy is dissipated in the form of heat energy minimally. The net electric flux density (D) can be expressed mathematically as follows:

$$D = \varepsilon_0 E + P \tag{1.64}$$

$$P = \varepsilon_0 \chi E \tag{1.65}$$

$$D = \varepsilon_0 (1 + \chi) E \tag{1.66}$$

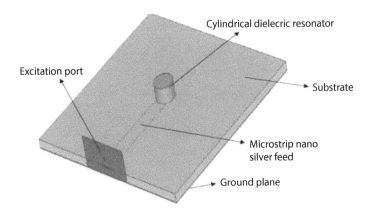

FIGURE 1.19   Nano DRA.

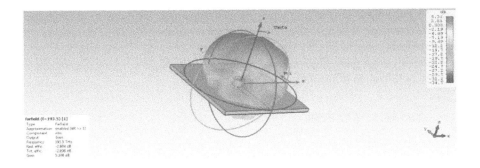

FIGURE 1.20   Nano DRA with microstrip feed.

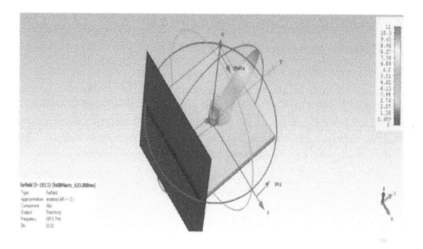

FIGURE 1.21 Nano DRA with silver nano feed.

where:
  P is the net polarization,
  E is the electric field intensity, and
  χ is the electric susceptibility.

This can be mathematically expressed as follows:

$$\varepsilon_r = 1 + \chi \qquad (1.67)$$

$$D = \varepsilon_0 \varepsilon_r E \qquad (1.68)$$

$$\varepsilon_r = \varepsilon_r' - j\varepsilon_r'' \qquad (1.69)$$

where ε_r' is the dielectric constant and $\tan \delta = \left(\frac{\varepsilon_r''}{\varepsilon_r'} \right)$ is called loss tangent or dissipation factor of dielectric.

Dielectric polarization occurs when the electromagnetic field passes through the dielectric material. It defines the dielectric property of the DR. The DR consists of the permanent electric dipoles which are randomly oriented. These randomly oriented dipoles align themselves in the direction of the field. This whole phenomenon is called polarization of the dielectrics.

1.13   QUALITY FACTOR

The Q-factor of the DRA can be written as given below:

$$Q = \frac{\omega W_S}{P_{\text{rad}}} \qquad (1.70)$$

where W_s is the energy stored and P_{rad} is the radiated power.

These quantities are determined by using (2.12):

$$W_s = \frac{\varepsilon_o \varepsilon_r bdh}{32}\left(1 + \frac{\sin(k_z h)}{k_z h}\right)\left(k_x^2 + k_y^2\right) \tag{1.71}$$

$$P_{rad} = 10k_o^4 |p_m|^2$$

where the magnetic dipole moment of the DRA is represented by p_m.

$$p_m = \frac{-j\omega 8\varepsilon_o (\varepsilon_r - 1)}{k_x k_y k_z} \tag{1.72}$$

The Q-factor can calculate the impedance bandwidth of the DRA by using (1.73).

$$BW = \frac{S-1}{Q\sqrt{S}} \tag{1.73}$$

where S is the VSWR.

The plots of the normalized Q-factor can be generated by using the above equations. The normalized Q-factor can be plotted as a function of **h/b** (DRA dimensions) for various values of **d/b** and dielectric constant.

The normalized Q-factor can be expressed mathematically as:

$$Q_e = \frac{Q}{\varepsilon_r^{3/2}} \tag{1.74}$$

Hence, these curves can approximate the Q-factor of DRA without depending upon the previous equations.

1.14 RESONANT MODES

The radiation characteristics of a DRA can be analyzed by using the resonant modes. The radiation phenomenon inside the DR is represented by using E- and H-field patterns, which determines the configuration of resonant modes. The field perturbation on the surface of the DRA produces the E-field distribution pattern, which defines the resonant modes. These modes can be TE, TM, and hybrid modes. The resonant modes can be calculated using the orthogonal Fourier basis functions. The resonant modes are responsible for determining the distribution of total current throughout the DRA to form the set of orthogonal solutions. The E_z and H_z fields are based on the concept of orthonormality (Van Bladel, 1975). Based on boundary conditions, the modes can be categorized as confined and non-confined modes. There are two conditions that define the confined and non-confined modes:

1. $\hat{n} \cdot E = 0$ and
2. $\hat{n} \times H = 0$.

Here, $\hat{n}$ is the unit vector normal to the surface of DR. The modes which follow both of these conditions are the confined modes and those which follow the first condition only are called non-confined modes (Mongia & Ittipiboon, 1997). The surface of evolution like cylindrical and spherical can have the confined modes only. The mode $TM_{01\delta}$ is the confined mode. The fundamental confined and non-confined modes act as the electric and magnetic dipoles.

The most common geometries of the DR are cylindrical and rectangular. In the case of cylindrical DR, transverse electric (TE), transverse magnetic (TM) and hybrid (HEM) modes exist. Some field distribution of the modes in the cylindrical DR have been computed and reported (Kajfez, Glisson, & James, 1984).

In case of the rectangular DRA, the existence of lower order TM modes was not verified experimentally (Mongia & Ittipiboon, 1997). Hence, the resonant frequency of the lower order TE modes is calculated using transcendental equations reported in the DWM (Marcatili, 1969).

The characteristic equation for the wave numbers k_x, k_y, and k_z for TE^y_{111} mode can be written as follows:

$$k_x a = m\pi - \tan^{-1}\left(\frac{k_x}{\epsilon_r k_{xo}}\right); \quad m = 1,2,3\ldots \tag{1.75}$$

$$k_{xo} = \left[(\epsilon_r - 1)k_o^2 - k_x^2\right]^{1/2} \tag{1.76}$$

$$k_y b = n\pi - \tan^{-1}\left(\frac{k_y}{\mu_r k_{yo}}\right); \quad \mu_r = 1, \ n = 1,2,3\ldots \tag{1.77}$$

$$k_{yo} = \left[(\epsilon_r - 1)k_o^2 - k_y^2\right]^{1/2} \tag{1.78}$$

$$k_z a = p\pi - \tan^{-1}\left(\frac{k_z}{\epsilon_r k_{zo}}\right); \quad p = 1,2,3\ldots \tag{1.79}$$

$$k_{zo} = \left[(\epsilon_r - 1)k_o^2 - k_z^2\right]^{1/2} \tag{1.80}$$

Here, k_{xo}, k_{yo}, and k_{zo} are the field decay factors outside the DRA and k_o is the free space wave number which can be determined using the following equation:

$$\epsilon_r k_0^2 = k_x^2 + k_y^2 + k_z^2 \tag{1.81}$$

$$k_0^2 = \omega_0^2 \mu_0 \epsilon_0 \tag{1.82}$$

Hence, we can calculate the resonant frequency by using the free-space wave number k_0.

The resonant frequency of an isolated rectangular DR in free space can be obtained as follows:

$$(f)_r \, m,n,p = \frac{c}{2\pi \sqrt{\mu\epsilon}} \sqrt{\left[\left(\frac{m\pi}{a}\right)^2 + \left(\frac{n\pi}{b}\right)^2 + \left(\frac{p\pi}{d}\right)^2\right]} \tag{1.83}$$

The height of the DR is considered double in the z-direction if the DR is placed above the ground plane.

1.14.1 THE CONCEPT OF SHORT MAGNETIC AND ELECTRIC DIPOLES

The concept of the short magnetic and electric dipole is necessary for understanding the resonant modes in DRA. A simple electric current carrying wire can be considered as the short electric dipole. The magnetic field always surrounds the wire that carries the electric current. In a similar fashion, the magnetic flux lines surrounded by the electric loops are considered as the magnetic dipole. In a practical scenario, the electric loop antenna is considered as the magnetic dipole.

The field configurations of the electric and magnetic dipoles are shown in Figure 1.22. It can be observed that electric as well as magnetic dipole radiates in the direction normal to its axis. Thus, a horizontal dipole will radiate in bore sight direction while the vertical dipole will radiate in the end-fire direction.

(a)   (b)

FIGURE 1.22 (a) Electric dipole and (b) magnetic dipole (solid line: electric field; dotted line: magnetic field).

The difference between electric and magnetic dipoles can be analyzed from the field configuration. It can be observed that there is a phase difference of 90° between E-field planes of electric and magnetic dipoles. The co-polarized component of the electric dipoles becomes the cross-polarized component of magnetic dipoles in a principal plane.

1.14.2 FEEDING TECHNIQUES OF DRA

Feeding techniques are used to couple the electromagnetic power inside the DR structure in many different ways. The Q-factor and resonant frequency have been greatly influenced by the coupling mechanism. There are various feeding techniques used in DRA structure (Petosa, 2007). The most commonly used feeding techniques in DRA are as follows.

1.14.2.1 Slot/Aperture Coupling

In this feeding mechanism, an aperture is made on the ground plane over which the DR structure is placed. To excite the magnetic field in the DRA, the magnetic current flows parallel along the slot length through aperture. The aperture consists of the slot and microstrip feed line beneath the ground plane. The most attractive feature of aperture coupling is that it offers the isolation of radiating aperture from unwanted spurious radiations as the feed network is below the ground plane. For impedance matching, the microstrip stub can be used which cancels out the reactive components of the slot. The slot in the DRA acts as the magnetic current source with the direction parallel to the slot length. The aperture should be in the strong magnetic field to achieve good coupling in the DRA. Due to easy fabrication, aperture coupling is the most common approach.

1.14.2.2 Coaxial Probe Feeding

The coaxial probe coupling technique involves the use of the center pin of the coaxial feeding line, which is extended from the ground plane to DR. The probe can be either adjacent or embedded inside the DR. The optimization of coupling and matching can be done by making the height adjustments of the probe inside the DRA. The probe can be considered as the vertical electric current source and can be located in a region of high E-fields in order to have strong coupling (Figures 1.23 through 1.27).

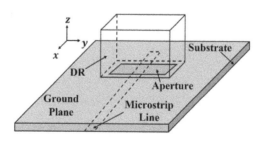

FIGURE 1.23 Aperture coupling in DRA.

(a)   (b)

FIGURE 1.24 Equivalent magnetic current for slot aperture. (a) Electric fields and (b) magnetic source equivalent.

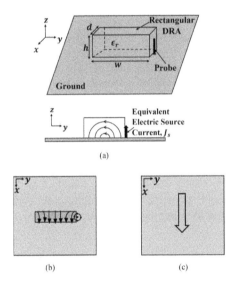

(a)

(b)   (c)

FIGURE 1.25 Probe coupling to a rectangular DRA. (a) Electric fields, (b) magnetic fields and (c) horizontal short magnetic dipole equivalence.

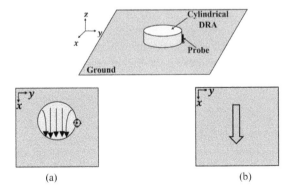

(a)   (b)

FIGURE 1.26 Probe coupling to the $HE_{11\delta}$ mode of the cylindrical DRA. (a) Magnetic fields $HE_{11\delta}$ mode and (b) horizontal short magnetic dipole equivalence.

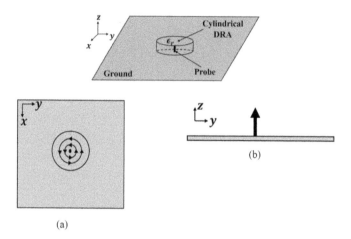

FIGURE 1.27 Probe coupling to the $TM_{01\delta}$ mode of the cylindrical DRA. (a) Magnetic fields ($TM_{01\delta}$ mode) and (b) vertical short electric monopole equivalence.

The modes can be excited by the optimization of probe location. For the rectangular DRA, the probe excites the $TE_{11\delta}^x$. For the cylindrical and split cylinder, the probe excites the $HE_{11\delta}$ and $TM_{01\delta}$ modes, respectively. To avoid the radiation from probe, its height should be less than the height of the DR. The major advantage of this feeding technique is that it can directly be matched to the 50Ω characteristic impedance. Hence, there is no need for the matching network. The coaxial probe is beneficial for the lower frequency ranges where aperture coupling is not possible practically.

1.14.2.3 Microstrip Line Feeding

This feeding technique is the simplest and most common method for DR coupling in microwave circuits. This coupling technique produces short horizontal magnetic dipoles, which excites the magnetic field inside the DRA. The dielectric constant of the DR material is the most dominant feature that affects the coupling. In this technique, a metallic strip is etched on one side of the substrate with definite thickness and permittivity. The substrate is grounded and metalized (Figure 1.28).

This coupling technique is easy to fabricate in comparison to other techniques. Strong coupling can be achieved using the higher permittivity of the DR. However, the maximum amount of coupling is reduced if the dielectric constant of the DR is reduced.

1.14.2.4 Antenna as *R, L, C* Circuit

The Figure 1.29 is obtained as an equivalent R, L, C circuit of an antenna.

$$Z = R + j\omega L - \frac{j}{wC}$$

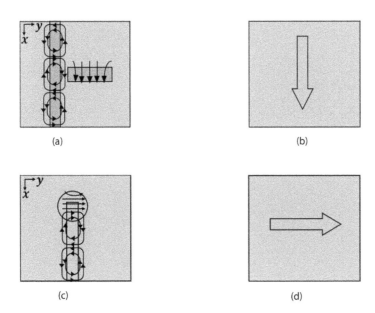

FIGURE 1.28 Fields and equivalent radiation models. (a) Magnetic fields, (b) short horizontal magnetic dipole equivalence, (c) magnetic fields, and (d) short horizontal magnetic dipole equivalence.

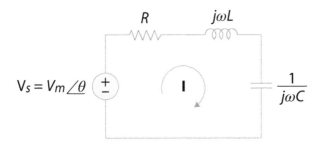

FIGURE 1.29 R, L, C circuit as an antenna.

Bandwidth at resonance:

Current at resonance is maximum.

$$i = \frac{V}{\sqrt{R^2 + \left(\omega L - \dfrac{1}{\omega C}\right)^2}} = \frac{i_{max}}{\sqrt{2}} = \frac{V}{R\sqrt{2}} :$$

Hence, $\sqrt{R^2 + \left(\omega_1 L - \dfrac{1}{\omega_1 C}\right)^2} = R\sqrt{2}$

$$R^2 + \left(\omega_1 L - \frac{1}{\omega_1 C} \right)^2 = 2R^2$$

Let $\omega_1 = \omega$

$$\Rightarrow \left(\omega L - \frac{1}{\omega C} \right)^2 = R^2$$

$$\Rightarrow \left(\omega L - \frac{1}{\omega C} \right)^2 = \pm R$$

$$\Rightarrow LC\omega^2 - RC\omega - 1 = 0$$

$$\Rightarrow \omega = \frac{RC \pm \sqrt{R^2 C^2 + 4LC}}{2LC}$$

$$\omega_H = \frac{RC + \sqrt{R^2 C^2 + 4LC}}{2LC}$$

Similarly taking $(-R)$ value,

$$\omega_L = \frac{-RC + \sqrt{R^2 C^2 + 4LC}}{2LC}$$

So, $BW = \omega_H - \omega_L = \dfrac{RC}{2LC} + \dfrac{RC}{2LC} = \dfrac{2RC}{2LC} = \dfrac{R}{L}$

So bandwidth $= \dfrac{R}{L}$ Radian/Sec

$$BW = \frac{R}{2\pi L} \text{ Hz}$$

1.15 CONCLUSION

Antenna works on the reciprocity principle. The parameter remains the same in trans and receives direction. Matching of input impedance and output impedance in an antenna is a very important part in antenna designing. Dimensions of microstrip line, slot, and DRA are frequency dependent similar to resonant frequency of antenna, which is also dependent on antenna dimensions and substrate material. A DRA has more advantages as compared to patch antennas due to aspect ratios and high range of permittivity of available materials. Switching from microwave frequency antenna design to optical antenna has great advantages of compact size and wide bandwidth. The drude antenna model is familiar for optical antennas. Nano dielectric resonator antennas are upcoming candidates for futuristic technologies of wireless communication due to compactness and ultra wide bands. Nanotechnology is presently in demand in medical sciences and wide band for 5G communications. Nano DRA operates at terahertz frequencies. Absorption coefficients, reflection coefficients and phase angel of biological tissue in different layers can be studied with a terahertz signal. It addresses nondestructive RF imaging by making use of electromagnetic radiation. This technology can provide high-resolution imaging on a real time basis. Terahertz antennas have large bandwidth, which is very useful for 5G applications. Non-linearity has been made use of in nano technologies. This is for offering high speed data such as 10 Gbps using next generation communication systems and antennas. Large bandwidth, greater capacity, increased security, lower latency and reliable connections of smart phone networks are the primary requirement for any reliable communication system. The continual development has been taking place toward enhancement of bandwidth, beam width control, making compact size antennas, obtaining high gain, introducing circular polarization and getting desired antenna radiation patterns. Different structures, feeding, and matching techniques have been developed to get optimum antenna parameters. There has been lot of developments in this field antenna, right from Yagi Uda antenna to the active shark fin type of antennas. These are used in practical applications in the field of entertainment, communication, surveillance, security, sensors, object identification, imaging, avionics integrated in radar systems, satellite systems, microwave communication systems, cars, airplanes, spacecraft, and mobile towers, etc.

REFERENCES

1. R. S. Yaduvanshi and H. Parthasarathy, *Rectangular DRA Theory and Design.* New Delhi: Springer, 2016.
2. A. Mehmood, O. H. Karabey, and R. Jakoby, "Dielectric Resonator Antenna with Tilted Beam," *IEEE Antennas Wirel. Propag. Lett.*, vol. 16, pp. 1119–1122, 2016.
3. A. Bonakdar and H. Mohseni, "Impact of Optical Antennas on Active Optoelectronic Devices," *Nanoscale*, vol. 6, no. 19, pp. 10961–10974, 2014.
4. M. Zou and J. Pan, "Investigation of Resonant Modes in Wideband Hybrid Omni directional Rectangular Dielectric Resonator Antenna," *IEEE Trans. Antennas Propag.*, vol. 63, no. 7, pp. 3272–3275, 2015.
5. Y. Zhang, A. A. Kishk, A. B. Yakovlev, and A. W. Glisson, "Analysis of Wideband Dielectric Resonator Antenna Arrays for Waveguide-Based Spatial Power Combining," *IEEE T. Microw. Theory*, vol. 55, no. 6, pp. 1332–1340, 2007.
6. S. Fakhte, "High Gain Rectangular Dielectric Resonator Antenna Using Uniaxial Material at Fundamental Mode," *IEEE Trans. Antennas Propagation*, vol. 65, no. 3, 2017.
7. G. Varshney, V. S. Pandey, R. S. Yaduvanshi, and L. Kumar, "Wide Band Circularly Polarized Dielectric Resonator Antenna with Stair-Shaped Slot Excitation," *IEEE Trans. Antennas Propag.*, vol. 65, no. 3, pp. 1380–1383, 2016.

8. G. Varshney, S. Gotra, V. S. Pandey, and R. S. Yaduvanshi, "Inverted-Sigmoid Shaped Multi-Band Dielectric Resonator Antenna with Dual Band Circular Polarization," *IEEE Trans. Antennas Propag*, vol. 66, no. 4, pp. 2067–2072, 2018.

9. G. Varshney, P. Praveen, R. S. Yaduvanshi, and V. S. Pandey, "Conical Shape Dielectric Resonator Antenna for Ultra Wide Band Applications," *International Conference on Computing, Communication & Automation*, pp. 1304–1307, 2015.

10. G. Varshney, V. S. Pandey, and R. S. Yaduvanshi, "Dual-Band Fan-Blade-Shaped Circularly Polarized Dielectric Resonator Antenna," *IET Microw. Antennas Propag*, vol. 11, no. 13, pp. 1868–1871, 2017.

11. A. Petosa and S. Thirakoune, "Rectangular Dielectric Resonator Antennas with Enhanced Gain," *IEEE Trans. Antennas Propag*, vol. 59, no. 4, pp. 1385–1389, 2011.

12. M. Singh, A. K. Gautam, R. S. Yaduvanshi, and A. Vaish, "An Investigation of Resonant Modes in Rectangular Dielectric Resonator Antenna Using Transcendental Equation," *Wireless Pers. Commun*, vol. 95, no. 3, pp. 2549–2559, 2017.

13. S. Gotra, G. Varshney, V. S. Pandey, and R. Yaduvanshi, "Dual-Band Circular Polarization Generation Technique with the Miniaturization of a Rectangular Dielectric Resonator Antenna," The Institution of Engineering and Technology, 2019.

2 Dielectric Resonator Antenna (DRA)

2.1 INTRODUCTION

Mobile phones, Wi-Fi, Bluetooth, WLAN, satellites, radar, sonar, automobiles, airplanes, walkie talkies, sensors, and wireless computer networks are used with mostly antennas and have become essential parts of human life. Working in the environment of 5G requires high speed and high data rate transmission. In present era aircrafts, autonomous vehicles are competent enough to make smart decisions. The main components of these wireless communications are **wideband antennas** and **multiband antennas** with specific polarization. In this chapter, we intend to focus on design techniques of antennas to provide extensive flexibility by incorporating **resonant mode theory**, circular polarization, control on polarization, gain control, **bandwidth control and beamwidth control** in DRA (dielectric resonator antenna) [1–15]. Dimensions of antenna have been taken based on operating and resonant frequency. These dimensions are called the physical lengths of an antenna. Fringing fields are taken into consideration to provide an account of antenna electrical lengths. Resonant frequency and mode excited are dimensions as well as structure dependent. Along with excitation input, excitation point in DRA also matters in mode excitation. Resonant modes form E-field patterns inside the antenna. E-field vectors can provide clarity on propagating fields in an antenna. Hence, antenna have a boundary value problem: transparent PMC (permanent magnetic walls) allow field to pass through. PEC (permanent electric walls) state does not allow E-fields to propagate. PEC walls reflect back electromagnetic waves. PMC walls are transparent and thus allow electromagnetic waves to pass through. This is also known as transparent state of DRA. It allows fields to pass through to space. Transmission lines and dielectric waveguide models can be developed for DRA. FEM and MOM are two other ways to obtain computational solutions of DRA. The resonant frequency of DRA (dielectric resonator antennas) is computed based on the DWM model.

DRA is an efficient radiator, can excite fundamental and higher order modes and has two different aspect ratios for resonant modes excitation and control. Different high permittivity DRAs are available in compact sizes. Permittivity range varies from 10 to 1600. The resonant frequency is inversely proportional to permittivity of DRA and dimensions. Hence, size and frequency design flexibility is available. DRA has high power handling capability. Hence, DRA has more advantages in design. Designing of conformal antennas using DRA is difficult task, hence it is a disadvantage of DRA.

The small frequency ratio is an important parameter in multiband antennas. Frequency ratio can predict separation in bands to have sufficient isolation in adjacent operating bands. Isolation between operating bands is a desired parameter to avoid cross talk or noise. Accurate computations of resonant frequency could be obtained using transcendental equations rather than characteristic equations.

DRA was proposed by Guillon and SA Long around 1980. Mongia et al. introduced DWM method in 1992. YMM Antar et al. introduced a modified wave guide model (MWGM) in 1998. DRA resonant frequency is a function of effective electrical dimensions of DRA.

Development of software tools like finite element method based HFSS or CST Microwave Studio has enabled computer aided design (CAD) models for antennas which can also predict eigen value-based resonance frequency. Use of transcendental equations for computing resonant frequency of DRA has reduced the error between computed and measured resonant frequency. The concept of higher modes has changed entire applications of the DRA domain. Later, modulation of modes can significantly introduce dimension minimization. Merging of higher order modes with fundamental modes introduced the concept of wide band DRAs. The equivalent circuit of DRA can be drawn with the help of R, L, C circuits and a solution of DRA can be had before design based on a mathematical background.

2.2 DRA PARAMETERS

Nano DRA has narrow beamwidth as compared to DRA. NDRA has wide bandwidth as compared to DRA. NDRA can be built with circular polarization. Circular polarization in an antenna has the advantage of creating robustness into transmitted signals to avoid noise and multi-path fading. DRA has multifold advantages as compared to traditional patch antenna. Nano DRA has further advantages of large bandwidth, non ionizing, and compact size (Figures 2.1 and 2.2).

DRA has the following antenna parameters:

FIGURE 2.1   Dielectric resonator antenna.

FIGURE 2.2   DRA fabricated with micro strip line.

2.2.1 RESONANT

Resonant frequency, input impedance, output impedance, radiated power, DRA gain, (reflection coefficient (s_{11})), VSWR, bandwidth, beam width, directivity, polarization (linear, elliptical, circular, LHCP, RHCP), substrate permittivity and permeability, dielectric constant, efficiency, quality factor, resonance modes, axial ratio, aspect ratio, frequency ratio, power gain, low profile, directive gain, dipole formation, dipole moments, fundamental and higher order modes, Poynting vector, power flux, magnetic vector potential and retarded potential, wave number, eigen value, boundary conditions (PEC and PMC), bessel function, green function, frequency spectrum, conductivity, $\lambda = \frac{C}{f}$, miniaturization and deminiaturization factor, group delay, loss tangent, Tan (∂_o), free space impedance $\left(\eta = \sqrt{\frac{\mu_o}{\epsilon_o}} = 377\Omega\right)$, permittivity $= 8.854 \times 10^{-12} / \frac{F}{m}$, permeability $= 4\pi \times 10^{-7}$ H/m, end fire radiation pattern, broadside radiation pattern, boresight radiation pattern, met material (negative permittivity material or near zero permittivity), multiband, channel capacity loss (**CCL**), MIMO, envelopment correlation coefficient (ECC), directive gain (DG), total active reflection coefficient (**TARC**) and mean effective gain (MEG) (Figures 2.3 through 2.8).

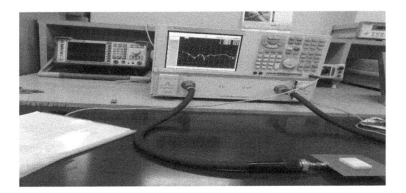

FIGURE 2.3 DRA measurements using VNA.

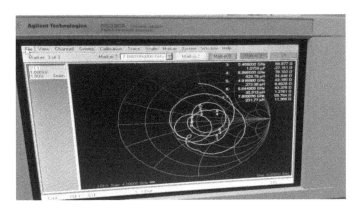

FIGURE 2.4 Smith chart showing Z_{11} measurements with VNA.

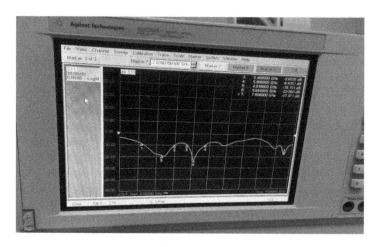

FIGURE 2.5   DRA showing S_{11} at different frequencies.

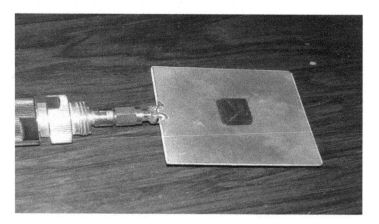

FIGURE 2.6   Sapphire DRA.

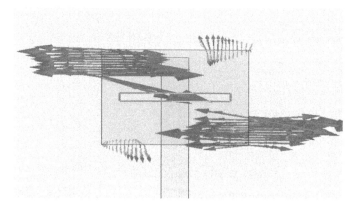

FIGURE 2.7   DRA fields propagation.

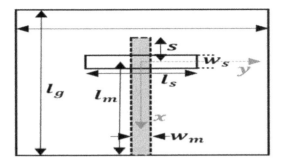

FIGURE 2.8   DRA microstrip line.

An antenna has several design parameters, such as, resonant frequency, reflection coefficient, radiation pattern, gain, polarization, beam width, bandwidth, frequency ratio, power ratio, directivity, azimuth angle, elevation angle, end fire beam/bore sight beam, broad side beam, circular polarization, linear polarization, elliptical polarization, fundamental mode, higher order mode, even modes, odd modes, S-parameters, cross polarization, co-polarization, isolation, envelop correlation coefficient, diversity gain, mean effective gain, received power, transmitted power, scattering parameters, one-port network, two-port network, fringing fields, surface current, short magnetic dipole, dipole moment, current density, retarded potential, electromagnetic wave, wave vector, PMC/PEC boundary conditions, permittivity, permeability, conductivity, radiations, wave function, Poynting vector, quality factor, second order differential equation, acceleration, deacceleration, hybrid mode, mode control, mode merging, near field and far field, anechoic chamber, hybrid antenna, DRA on Patch (DROP), fields inside DRA, axial ratio, axial ratio bandwidth, impedance bandwidth, coupling of fields, pattern diversity, link budget equation, etc.

Ground plane, substrate type and DR material, slot, micro strip feed, stub and SMA connector with input excitation gives the design shape of a DRA at a specific frequency and impedance for radiation. Dielectric boundary conditions obey the continuity equation to introduce acceleration into charges so as to form radiation.

2.3   DESIGN OF RECTANGULAR DRA

The rectangular DRA design process is very simple. One can very easily fabricate and measure the results using VNA. Any antenna design can be verified using mathematical formulation for design frequency. It can be verified using simulation results. This is again cross verified using experimentation. Hence if all three results are same then the designer feels confident about the target design. Following DRA design steps are given:

$$\text{Wavelength of antenna}\left(\lambda\right) = \frac{c\,(\text{Velocity of light})}{f\,(\text{Frequency of antenna})}$$

Dimensions of DRA proportional to $= \dfrac{\lambda_0}{\sqrt{\epsilon r}}$;

where λ_0 is the free space wavelength and ϵ is the permittivity of DR material used.

Slot dimensions:

$$(\text{ls}) \text{ length of slot} = \dfrac{0.4\lambda_o}{\sqrt{\epsilon r}}$$

(ws) width of slot $= 0.2$ ls

Resonant frequency of rectangular DRA:

$$fr_{\text{mnp}} = \dfrac{c}{2\pi\sqrt{\epsilon r}} \sqrt{\left(\dfrac{m\pi}{a}\right)^2 + \left(\dfrac{n\pi}{b}\right)^2 + \left(\dfrac{p\pi}{2d}\right)^2} \; ;$$

$$\epsilon r_{\text{eff}} = \left(\dfrac{\lambda_o}{\lambda_g}\right)^2 \; ;$$

The value of aspect ratio must be chosen between 0.5 to 2.5 for better results.

DRA can be realized using the R, L, C circuit method by converting its equivalent circuit. DRA equivalent circuit can be converted into the R, L, C series circuits for obtaining the solution of resonant frequency.

The radiation pattern can be changed by exciting different modes. Electrical boundaries along with excitation points can excite resonant modes inside DRA. Resonant modes can provide physical insight of the antenna working. The amount of coupling is reduced if the dielectric permittivity of the DRA is low. DRA resonant frequency decreases with increase of dielectric permittivity. DRA has many fold advantages as compared to patch antennas. DRAs have simple geometries like hemisphere, cylindrical, rectangular, conical, etc. They are suitable for wireless communication. The characteristics of DRAs are determined by their shape, size, type of dielectric material, and excitation.

2.4   DRA USED IN 5G

Nano DRAs are used for 5G communication because of large bandwidth at terahertz. Higher order modes are generated in DRA and they are merged with fundamental modes to get wide band response.

2.4.1   CHANNEL CAPACITY

Channel capacity can achieve a very high data rate (can reach up to 10 Gbps in 5G).

$$C = B \cdot \text{Log2} \; (1 + \text{SNR})$$

where:

C is the channel capacity and
B is the bandwidth of the channel.

3.1 to 10.6 GHz bandwidth antennas are classified as wideband antennas.

2.4.2 Resonant Frequency

$$f_r = \frac{c}{2\pi\sqrt{\varepsilon_r}}\sqrt{k_x^2 + k_y^2 + k_z^2} \tag{2.1}$$

Wave number k_x, k_y and k_z are in x, y and z directions and ε_r is the permittivity of DRA and c is the velocity of EM wave in free space. The relative permittivity (ε_r) is an important parameter because it affects the size, bandwidth, quality factor, and resonant frequency of the DRA. The DRA size depends on $\lambda_0/\sqrt{\varepsilon_r}$ factor, where λ_0 is the wavelength at resonant frequency. Hence, ε_r also affects the size of DRA significantly.

Q-factor of a resonator is defined as

$$Q = 2\omega_0 \frac{\text{Stored energy}}{\text{Radiated power}} \tag{2.2}$$

where loss tangent (tanδ) is dissipation of the electrical energy due to electrical conductions, dielectric relaxation, dielectric resonance, and loss from non-linear processes.

Bandwidth in terms of voltage standing wave ration (VSWR) and quality factor (Q) of antenna as,

$$BW = \frac{VSWR-1}{Q\sqrt{VSWR}} \tag{2.3}$$

The VSWR can be defined in terms of the reflection coefficient $|\Gamma|$ as:

$$VSWR = \frac{1+|\Gamma|}{1-|\Gamma|}$$

The reflection coefficient is also known as S_{11} or return loss of antenna.

2.4.3 Polarization

Polarization in antenna is defined as the orientation of the electric field, in the direction of propagation. The plane containing the electric and magnetic fields is called the plane of polarization. The polarization of an antenna with reference to far-field radiated fields in the direction of the propagation is defined by axial ratio of the fields. The polarization quality is expressed by the axial ratio. The desired polarization of the wave to be radiated by the antenna is called co-pol and cross pol. Getting circular polarization can be good for communication as it extends robustness to trans and receive signals. Getting circular polarization into three band antenna is a problem in research today.

Figure 2.9 is showing impedance bandwidth. Axial ratio band width defines polarization achieved.

$$\boldsymbol{RL[dB]} = 10 \times \log\left(|\Gamma|^2\right)$$

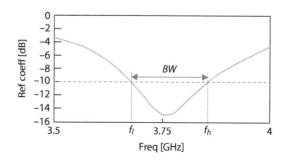

FIGURE 2.9 S_{11} of DRA.

$$-10 = 10\log\left(\frac{P_r}{P_i}\right) \quad RL = |\Gamma|^2 = \frac{\langle P_r \rangle}{<P_i>}$$

$$P_r = 0.1 \times P_i$$

2.4.4 DIRECTIVITY

Directivity is often calculated by taking the ratio of the maximum power density (P_{max}) to the average power density taken the whole sphere, i.e.,

$$D = \frac{P_{max}}{\dfrac{1}{4\pi}\displaystyle\int_{\theta=0}^{\pi}\int_{\theta=0}^{\pi} P(\theta,\varphi)\sin\theta d\theta d\varphi} \tag{2.4}$$

Directivity = $10 \log_{10} D$ dB (dBi)

Maximum gain (G) of an antenna in a particular direction is called directivity.

2.5 ELECTRICAL LENGTH OF DRA

There is a difference between physical length and electrical length of an antenna. Electrical length is larger than physical length. This electrical length increase is due to fringing effect (Figure 2.10).

The transcendental equation provides accurate resonant frequency and it is given as follows:

$$k_y \tan\left(\frac{k_y d}{2}\right) = \sqrt{(\epsilon_r - 1)k_0^2 - k_y^2} \tag{2.5}$$

However, the above equation can also be written as follows:

$$k_y b + 2 \tan^{-1}\left(\frac{k_y}{k_{y0}}\right) = n\pi \tag{2.6}$$

where $k_{y0} = \sqrt{(\epsilon_r - 1)k_0^2 - k_y^2}$

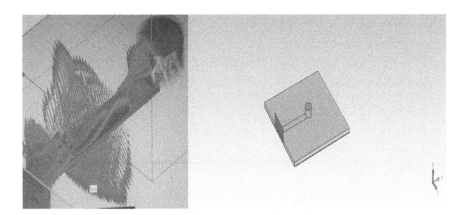

FIGURE 2.10 Excited state of nano DRA.

By binomial expansion of (2.6), we can obtain:

$$k_y b + 2 k_y \left(\frac{\sqrt{(\epsilon_r - 1)k_0^2 - k_y^2}}{(\epsilon_r - 1)k_0^2} \right) = n\pi \qquad (2.7)$$

Rearranging the above variables, we get:

$$k_y = \frac{n\pi}{b \left[1 + 2 \left(\dfrac{1}{bk_0 \sqrt{(\epsilon_r - 1)}} \right) \right]} \qquad (2.8)$$

$k_y = \dfrac{n\pi}{\hat{b}}$, where $\hat{b}$ is the effective width.

The relation between b (physical width) and $\hat{b}$ (effective width) can be expressed as given below:

$$\hat{b} = b + \left(\frac{2}{k_0 \sqrt{(\epsilon_r - 1)}} \right) \qquad (2.9)$$

In TE_{mnp}^y mode, the m, n and p are integers for half wave variations in x, y and z direction. In $TE_{1\delta 1}^\psi$ mode, half wave variation is in the y-direction but the integer is less than 1. The parameter δ is a fraction of a half wave variation in the y-direction with a value between 0 and 1.

$$\delta = \frac{k_y}{(\pi/b)} \qquad (2.10)$$

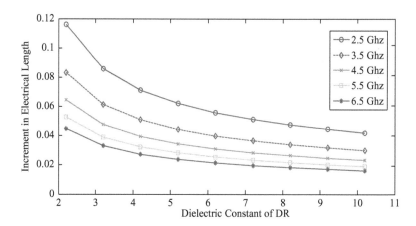

FIGURE 2.11   Effective width and dielectric constant of DRA.

From (2.8) and (2.10), δ can be derived as follows (Figure 2.11):

$$\delta = \frac{1}{\left(1 + 2\left(\dfrac{1}{bk_0\sqrt{(\epsilon_r - 1)}}\right)\right)} \qquad (2.11)$$

2.6   RESONANT MODES CONTROL IN DRA

This introduces us to resonant modes and their control. The generation of fundamental and higher modes in DRA has become possible. These modes can be shifted, mixed or merged. Mutilple modes can be excited in one DRA. Design can provide flexibility to operate at a higher order modes. Also, as compared to fundamental mode, higher order modes can provide high gain along with miniaturized design. Two adjacent modes can be merged into a single mode for bandwidth enhancements. Two or three modes generated can be shifted to different resonant frequencies to operate them into multi mode configurations. Hence, DRA designs extend flexibility to designer to operate same antenna for different applications with a control on resonant modes excited into DRA. $\mathbf{TE}^y_{mnp}$ Means field is propagating in the y direction, and mnp are different combinations of higher order modes. Also, the y dimension is shorter than x and z in the DRA. Figure 2.1 is a magnetic dipole formation used to represent mode excitation. These short magnetic dipoles will be formed depending on the electrical length of DRA. The number of short magnetic dipoles may vary depending on the excitation and wavelength of DRA (Figure 2.12).

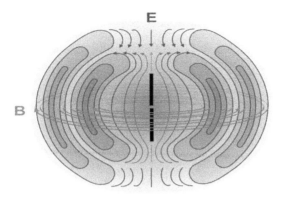

FIGURE 2.12   Formation of short magnetic dipoles in DRA.

2.6.1   DIPOLE MOMENT

Dipole moment equation is given below

$$d = \sum e \cdot r$$

where:
   d is the dipole moment,
   e is the charge, and
   r is the distance between two charges.

$$d = \frac{d}{dt} \sum e \cdot r = \sum e \cdot v$$

$$\ddot{d} = \frac{d}{dt} \sum e \cdot v$$

The charges can radiate only under acceleration condition. There will be no radiation with fixed or uniform velocity.

2.6.2   RESONANT MODES MATHEMATICAL DESCRIPTION

$$\left(\nabla^2 + \frac{\omega^2}{c^2} \right) \left(\begin{matrix} E_z \\ H_z \end{matrix} \right) = 0$$

The discrete resonant modes $\omega(mnp)$, at output of L, C oscillators with different L, C values. $\underline{E}$ and $\underline{H}$ fields formats are different inside the resonator cavity. The superposition of four, three vector valued basis functions for easy understanding of resonating modes inside DRA are given as follows:

$$\underline{E}(x,y,z,t) = \sum_{m,n,p=1}^{\infty} Re\left\{c(mnp)e^{j\omega(mnp)t}\underline{\psi}_{mnp}^{E}(x,y,z)\right\}$$

$$+ \sum_{m,n,p=1}^{\infty} Re\left\{d(mnp)e^{j\omega(mnp)t}\underline{\bar{\phi}}_{mnp}^{E}(x,y,z)\right\}$$

and

$$\underline{H}(x,y,z,t) = \sum_{m,n,p=1}^{\infty} Re\left\{c(mnp)e^{j\omega(mnp)t}\underline{\psi}_{mnp}^{H}(x,y,z)\right\}$$

$$+ \sum_{m,n,p=1}^{\infty} Re\left\{d(mnp)e^{j\omega(mnp)t}\underline{\bar{\phi}}_{mnp}^{H}(x,y,z)\right\}$$

Here, $\{C(mnp)\}$ and $\{d(mnp)\}$ are linear combinations of coefficients.

2.6.3 THE SOLUTION OF RESONANT MODES

1. H_z, E_z fields are expressed as $u_{mnp}(x,y,z)\,v_{mnp}(x,y,z)$. Also, ω_{mnp} is given by solving Maxwell's equations with given boundary conditions.
2. At $z = 0$; surface (x, y) current density is given as

$$\left\{J_{sX}(x,y,t),\ J_{sY}(x,y,t)\right\}$$

3. Surface current density is equated with magnetic fields

$$\left\{J_s(x,\ y,\ \delta) = \left(J_{sX},\ J_{sy}\right) = \hat{z}\times\underline{H}) = \left(-H_y,\ H_x\right)\right\};$$

at $z = 0$; amplitude coefficients are obtained on expansion of H_z, termed $d_{(mnp)}$, and Ez termed C_{mnp}.
4. When. $E_y\big|_{z=0}$ to zero, we compute the coefficients $d_{(mnp)}$ for Hz and C_{mnp} of E_z.
5. Excited by ωmnp and arbitrary feed position in the xy plane $(x_o,\ y_o)(\phi_o,\ \theta_o)$

$$H_\perp = \left[\sum_{mnp} R_e \left\{ \tilde{d}_{(mnp)} e^{j\omega(mnp)t} \right\} \nabla_\perp \tilde{u}_{mnp}(x,y,z) \right]$$

$$- \sum_{mnp} R_e \left\{ \tilde{C}_{(mnp)} e^{j\omega(mnp)t} \right\} \nabla_\perp \tilde{u}_{mnp}(x,y,z).$$

and similarly $E_\perp$. boundary conditions to be developed.

The obtained final results are: if all four walls are PMC and top and bottom walls are PEC, then we get the results as follows:

$$u_{mnp} = \sin\sin\sin = E_z$$

$$v_{mnp} = \cos\cos\cos = H_z$$

2.6.4 DRA Modes Propagation Parameters

The most important parameters are: excitation, coupling, medium, dimensions, point of excitation and input impedance (Figures 2.13 and 2.14).

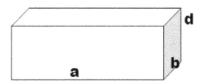

FIGURE 2.13 Rectangular DRA with a feed probe inserted in the *xy* plane.

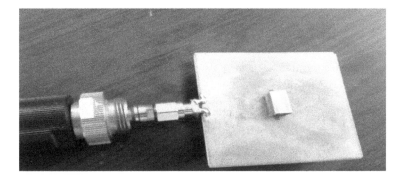

FIGURE 2.14 Rectangular DRA under excitation stage.

2.6.4.1 Theory of Current Excitation into DRA

DRA has a, b and d dimensions with probe feed under boundary conditions that can be solved for radiated fields as given below:

$E_X, E_y = 0$, top and bottom plane being electric walls.
$H_X = 0$, side walls being magnetic walls.

$H_z(x,y,z,t) =$

$$\sum_{mnp} C\,(m,n,p)\frac{2}{\sqrt{ab}}\sin\left(\frac{m\pi x}{a}\right)\sin\left(\frac{n\pi y}{b}\right)\sin\left(\frac{p\pi z}{d}\right)\mathrm{Cos}\,(\omega(m,np)t + \phi\,(m,np))$$

Fourier basis function $= \dfrac{2}{\sqrt{ab}}\sin\left(\dfrac{m\pi x}{a}\right)\sin\left(\dfrac{n\pi y}{b}\right) = u_{mn}(x,y)$ for convenience.

$$E_z(x,y,z,t) = \sum_{mnp} D(m,n,p)\frac{2}{\sqrt{ab}}\cos\left(\frac{m\pi x}{a}\right)\cos\left(\frac{n\pi y}{b}\right)\cos\left(\frac{p\pi z}{d}\right)$$

$$\{\cos\,(\omega(m,np)t + \psi(m,np)\}$$

Let, $\dfrac{2}{\sqrt{ab}}\cos\left(\dfrac{m\pi x}{a}\right)\cos\left(\dfrac{n\pi y}{b}\right) = v_{mn}(x,y)$ for convenience $=$ Fourier basis function.

$E_z = -j\omega A_z - \dfrac{\partial\overline{\phi}}{\partial z}$; Lorentz Gauge conditions

A_z can be expressed as:

$$\hat{A}_z(x,y,z,\omega) = \frac{\mu}{4\pi}\hat{I}(\omega)\frac{\delta l\,e^{-j\frac{\omega r}{c}}}{r}; \text{ where } \delta l = \text{probe length}$$

$\mathrm{Div}\,\hat{\underline{A}} = \dfrac{\partial \hat{A}_z}{\partial z}$; need to be computed.

Probe-based propagation fields are:

$$\frac{l_2}{2} < \left|x - \frac{a}{2}\right| < \frac{a}{2}, \quad \frac{l_2}{2} < \left|y - \frac{b}{2}\right| < \frac{b}{2}$$

The magnetic vector potential:

$$\frac{\partial \hat{A}_z}{\partial z} = \frac{\mu \hat{I}\delta l}{4\pi}\left[\frac{\partial}{\partial z}\frac{e^{-jkr}}{r}\right] = \frac{\mu \hat{I}\delta l}{4\pi}\left(\frac{\cos\theta}{r^2} - \frac{jk\cos\theta}{r}\right)e^{-jkr} = -\frac{j\omega}{c^2}\hat{\phi}$$

And scalar potential will be,

$$\hat{\underline{\phi}} = \frac{\mu \hat{I} \delta l j c^2}{4.\pi\,\omega} \cos\theta \left(\frac{1}{r^2} - \frac{jk}{r} \right) e^{-jkr} = \frac{\mu \hat{I} \delta l c^2}{4.\pi\,\omega} \left(\frac{z}{r^3} - \frac{jkz}{r^2} \right) e^{-jkr}$$

Differentiating $\hat{\underline{\phi}}$ w.r.t. z,

$$\frac{\partial \hat{\underline{\phi}}}{\partial z} = \frac{\mu \hat{I} \delta l c^2}{4\pi\,\omega} \left(\frac{1}{r^3} + \frac{3z^2}{r^5} - \frac{jk}{r^2} - \frac{2jkz^2}{r^4} + \left(\frac{z}{r^3} - \frac{jk\,z}{r^2} \right) \left(\frac{-jkz}{r} \right) \right) e^{-jkr} \qquad (2.12)$$

When, $\hat{E}_z = -j\omega\,\hat{A}_z - \dfrac{\partial \hat{\underline{\phi}}}{\partial z}$, substituting $\dfrac{\partial \hat{\underline{\phi}}}{\partial z}$ in $\hat{E}_z$,

$$\hat{E}_z = \left[\frac{-j\omega\mu \hat{I}\delta l}{4\pi\,r} - \frac{\mu \hat{I}\delta l c^2}{4\pi\,\omega} \left(\frac{1}{r^3} + \frac{3z^2}{r^5} - \frac{jk}{r^2} - \frac{2jkz^2}{r^4} - \frac{jkz^2}{r^4} - \frac{k^2 z^2}{r^3} \right) \right] e^{-jkr} \qquad (2.13)$$

If we take $\theta = \dfrac{\pi}{2}$, $z = 0$,

$$\hat{E}_z = \frac{\omega c \mu \hat{I}\delta l}{4\pi\,k} \left[\frac{-jk^2}{r} - \left(\frac{1}{r^3} + \frac{3\delta^2}{r^5} - \frac{jk}{r^2} - \frac{2jk\delta^2}{r^4} - \frac{jk\delta^2}{r^4} - \frac{k^2\delta^2}{r^3} \right) \right]$$

Also, $r = \sqrt{x^2 + y^2 + \delta^2}$

$\dfrac{k^2 r^2}{r} = \dfrac{1}{r}$, for $r = \lambda = 2\pi/k$;

For inductive zone $r^2 \approx \delta^2$
given that, $\delta \ll r$
Minimum of $r \approx l_2$ and maximum of $r = (a, b)$;
$kr \ll 1$

Hence, $\hat{E}_z \approx \dfrac{\mu c \hat{I}\delta l}{4\pi\,kr^3}$ and $\approx \dfrac{\mu c^2 \delta l}{4\pi\,r^3\,j\omega}\,\hat{I}(\omega)$;

$$E_z(t, x, y\delta) \approx \frac{\delta l}{4\pi\epsilon \left(\sqrt{x^2 + y^2} \right)^3} \int_0^t I(\tau)\,d\tau \approx \frac{Q(t)\delta l}{4\pi\epsilon\,r^3} \qquad (2.14)$$

Poynting vector is defined as a radiated power flux per unit solid angle or power radiated in a particular direction in a specified angular zone.

$$H = \nabla \times A$$

$E = -\nabla \varnothing - \dfrac{dA}{dt}$; scalar and magnetic vector potential from Lorentz gauge conditions.

$S = \left(E \times H^*\right)$; S is Poynting vector (energy flow or flux).

$Z = \dfrac{P_{rad}}{|I|^2}$ = input impedance

Radiation pattern due to probe current I (t) and probe length dl in DRA:

$\dfrac{\mu I \, \vec{dl} \, e^{-jkr}}{4\pi r} = \vec{A}$; where A is magnetic vector potential

From Helmholtz equation $\left|\vec{A}\right|$

$$\underline{E} = -j\omega \underline{\vec{A}}$$

Radiated power can be given as:

$$\dfrac{\left|\underline{E}\right|^2}{2\eta} = \dfrac{\omega^2 \left|\vec{A}\right|^2}{2\eta} \qquad \sqrt{\dfrac{\mu}{\epsilon}} = \eta = \text{characteristic impedance.}$$

$$\underline{\vec{A}} = \dfrac{\mu}{4\pi} \int\limits_{Volume} \dfrac{J\left(\underline{r}',\omega\right) e^{-jk\left|\underline{r}-\underline{r}'\right|}}{\left|\underline{r}-\underline{r}'\right|} d^3r' \; ; \text{ at source.}$$

We know that radiation pattern can be defined by electrical field intensity E_θ, E_ϕ:

$$E_\theta = -j\omega A_\theta \; \text{ and } \; \underline{A}_\theta = \hat{\theta}.\underline{A}$$

Antenna surface current density can be expressed as:

$$J\left(\underline{r}',\omega\right) = \sum\limits_{mnp} \underline{J}_s\left[mnp,\underline{r}'\right] e^{j\omega(mnp)t}; \text{ where, } r = (x,\, y,\, z)$$

The magnetic vector potential in terms of J can be written as:

$$\underline{A} = \dfrac{\mu}{4\pi} \sum\limits_{mnp} \int \dfrac{\underline{J}_s\left[mnp,\,\underline{r}'\right] e^{j\omega(mnp)\left(t - \frac{\left|\underline{r}-\underline{r}'\right|}{c}\right)}}{\left|\underline{r}-\underline{r}'\right|} ds\left(\underline{r}'\right); \text{ where } ds \text{ is the surface of RDRA}$$

$$= \frac{\mu}{4\pi} \frac{e^{jkn}}{|\underline{r} - \underline{r}'|} \sum_{mnp} \int_s \underline{J}_s \left[mnp, \underline{r}' \right] e^{j\omega(mnp)\hat{r}.\underline{r}'} \, ds\left(\underline{r}' \right)$$

$$H_\phi = E_\theta / \eta, \quad H_\theta = -\frac{E_\phi}{\eta}.$$

Hence, radiated power can be given as:

$$P_{rad} = \frac{1}{2\eta} \left(|E_\theta|^2 + |E_\phi|^2 \right)$$

$$E_\theta = \frac{\mu}{4\pi r^2} R_e \sum_{mnp} \int_s \left\{ J_{sx} \left[mnp, \underline{r}' \right] \cos\varphi \cos\theta + J_{S_y\left[mnp, \underline{r}' \right]} \sin\phi \cos\theta - J_{Sz\left[mnp, \underline{r}' \right]} \sin\theta \right\}$$

$$\exp\left(j\omega \frac{(mnp)}{c} \right) \left\{ \left(x' \cos\phi \sin\theta + y' \sin\phi \sin\theta + z' \cos\theta \right) ds\left(\underline{r}' \right) \right] e^{j\omega(mnp)t};$$

$$E_\varphi = R_e \sum_{mnp} \int_s \left\{ -J_{sx}\left(mnp, \underline{r}' \right) \sin\phi + J_{S_y\left[mnp, \underline{r}' \right]} \cos\phi \right\} e^{j\frac{\omega(mnp)}{c}} \left(x' \cos\varphi \sin\theta \right.$$

$$\left. + y' \sin\phi \sin\theta + z' \cos\theta \right) ds\left(\underline{r}' \right) e^{j\omega(mnp)t} \, ds\left(\underline{r}' \right);$$

Radiated power P_{rad} x, y, z component-wise can thus be defined as:

$$P_x[\hat{r} \,|\, mnp] = \int_s J_{sx}\left(mnp, \underline{r}' \right) e^{j\frac{\omega(mnp)\hat{r}.\underline{r}'}{c}} \, ds\left(\underline{r}' \right)$$

$$P_y[\hat{r} \,|\, mnp] = \int_c J_{s_y}\left(mnp, \underline{r}' \right) e^{j\frac{\omega(mnp)\hat{r}\underline{r}'}{c}} \, ds\left(\underline{r}' \right)$$

$$P_z[\hat{r} \,|\, mnp] = \int_s J_{sz}\left(mnp, \underline{r}' \right) e^{j\frac{\omega(mnp)\underline{r}.\underline{r}'}{c}} \, ds\left(\underline{r}' \right) \qquad (2.15)$$

The Radiation Pattern

Now, power radiation pattern can be defined as:

$$\frac{|E_\theta|^2+|E_\phi|^2}{2_\eta} = \frac{1}{2}\left\{\sum_s\left(E_{s_\theta}\,e^{j\omega(s)t}+E_{s_\phi}\,e^{j\omega(s)t}\right)\right\}\times\frac{1}{2}\left\{\left(\sum_s H_{s_\theta}\,e^{j\omega(s)t}+\sum_s H_{s_\phi}\,e^{j\omega(s)t}\right)\right\}$$

$$=\frac{1}{4}\left(\sum_s E_s\times H_m^*\,e^{j(\omega_s-\omega_m)t}+\sum_s E_s^*\times H_m\,e^{j(\omega_m-\omega_s)t}\right)$$

$$=\frac{1}{4}\sum_s\left[E_{s\theta}\times H_{s\varnothing}^*+E_{s\varnothing}^*\times H_{s\theta}\right]$$

$$=\frac{1}{2}R_e\sum_s\left(E_{s\theta}\times H_{s\varnothing}^*\right)$$

$$=\frac{1}{2}\left(E_{s\theta}\,\hat\theta+E_{\varnothing}\,\widehat{\varnothing}\;\right)\times\left(\frac{E_{s\theta}^*}{\eta}\,\hat\varphi-\frac{E_{s\varnothing}}{\eta}\,\hat\theta\right)$$

$$=\sum_s\frac{|E_{s\theta}|2}{2\eta}\,\hat r+\frac{|E_{s\varnothing}|2}{2\eta}\,\hat r;$$

2.7   DRA EQUIVALENT CIRCUIT

The Figure 2.15 is equivalent circuit of DRA.

2.7.1   DRA Equivalent Circuit Solution

$$Y_D=\frac{1}{R_D}+\frac{1}{s\,L_D}+s\,c_D \qquad y_s=\frac{1}{R_s}+\frac{1}{s\,L_s}+s\,c_S$$

$$Z_{in\,DRA}=\frac{1}{n^2}Z_D=\frac{1}{n^2}\frac{1}{y_D}=\frac{1}{n^2\left(\dfrac{1}{R_s}+\dfrac{1}{s\,L_s}+S\,c_S\right)}$$

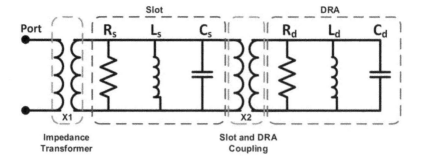

FIGURE 2.15   DRA equivalent R, L, C circuit.

$$Z_{inDRA} = \frac{s\,L_s\,R_s}{n^2\left(R_s + s\,L_s + s^2 L_s C_s R_s\right)}$$

$$Z_s = \frac{1}{y_s} + Z_{inD}$$

$$Z_s = \frac{1}{\dfrac{1}{R_s} + \dfrac{1}{s\,L_s} + S\,c_S} + Z_{in}$$

$$Z_s = \frac{R_s\,s\,L_s}{s^2 L_s C_s R_s + R_s + S\,L_s} + \frac{S L_D R_D}{n^2\left(R_D + S L_D + S^2\,L_D C_D R_D\right)}$$

$$Z_{inDRA} = R_s L_s s\left\{n^2\left(s^2 L_D R_D C_D + s L_D + R_D\right)\right\} +$$

$$s L_D\,R_D\left(s^2 L_s R_s C_s s L_s\,R_s\right) / n^2\left(s^2 L_D R_D C_D + s L_D + R_D\right)\left(s^2 L_s R_s C_s + s L_s + R_s\right)$$

$$Z_{in} = \frac{1}{n^2}\,Z_{inDRA}$$

$$Z_{in} = \frac{1}{n^4}\Big[n^2 s^3 L_D L_s R_s\,R_D\,C_D + n^2\,s^2 L_D L_s R_s + s R_s\,R_D\,L_s + s^3 L_D R_D L_s R_s C_s$$

$$+ s^2 L_D R_D L_s + s L_D R_D R_s$$

$$\left(s^4 L_D\,R_D\,C_D L_s R_s\,C_s\right) + s^2\left(L_D\,R_D\,C_D R_D + L_s R_s\,C_s R_S\right)$$

$$+ S^3\left(L_D R_D C_D L_s + L_D L_s R_s C_s\right) + R_s R_D + s\left(L_D R_s + L_s R_D\right)\Big]$$

$$Z_{in} = \frac{1}{n^4}\Big[s^3\left(n^2 L_D R_D C_D R_s L_s + L_D R_D L_s R_s\,C_s\right)\Big] + s^2\left(n^2 L_D L_s R_s + L_D R_D R_s\right)$$

$$+ s\left(L_D R_D R_s + R_s R_D L_s\right)$$

$$\left(s^4 L_D\,R_D\,C_D L_s R_s\,C_s + s^3\left(L_D R_D C_D L_s + L_D L_s R_s\,C_s\right) + s^2\left(L_D R_D C_D R_D + L_s R_s C_s R_s\right.\right.$$

$$+ S\left(L_D R_s + L_s R_D\right) + R_s R_D$$

2.7.2 DRA Radiation Mechanism

Radiation in DRA is introduced due to excitation. DRA material permittivity ϵ_r forms capacitance, hence charge can define energy stored into DRA. The field pattern into the bounded region is created due to the reflection and refraction phenomenon because of excitation. The field will pass through transparent walls, penetrating through the dielectric medium due to the fringing effect at the side wall, and energy radiates into medium due to transparent walls (PMC).

$$n \times H = 0 \qquad H \quad \text{Tangential}$$
$$n \cdot E = 0 \qquad E \quad \text{Normal}$$
$$\text{PMC}$$

$$n \times E = 0 \qquad E \quad \text{Tangential}$$
$$n \cdot H = 0 \qquad H \quad \text{Normal}$$
$$\text{PEC}$$

Emission of this radiation is due to oscillatory currents given to DRA as excitation. Each component can be separately obtained by separation of variables for various studies. No generality is lost by the system in terms of potential, fields and radiation from systems of charges and currents to power radiated into space.

Power radiated per unit solid angle (energy per unit area per unit time) forms the beam.

$$\left. \begin{aligned} H &= n \times E \\ E &= H \times n \end{aligned} \right\} \quad \text{by duality}$$

Relative amplitude of a wave can be expressed by $d(mnp)$ and $c(mnp)$, the solution of λ, k, ψ, the wavelength, wave vector and wave functions. Hence, the Poynting vector gives energy flow as follows:

$$S = c \cdot \frac{H^2}{4\pi} \cdot n$$

Current densities can be compared (i.e. Fourier components $d(mnp)$, H_z and $c(mnp), E_z$) by probe excitation currents. Here $E_\perp$ and $H_\perp$ can be worked for separation variables to obtain E_x, E_y and H_x, H_y fields.

$$h_{mn}^2 = \gamma^2 + h^2$$

$$\text{where} \qquad \gamma = \frac{j\pi p}{d}$$

$$H_z = v_{mnp} = v_{mn}(x,y)e^{-\frac{j\pi pz}{d}}$$

Since permittivity of material inside DRA is much higher than air presents outside the medium, creating dielectric – air interface, can be said to be permanent magnetic conducting (PMC) walls for e.m.waves. Cavity acts like LC tuned circuit. Thus it resonates at microwave frequency. These electromagnetic waves bounce back and forth between PEC walls due to reflection form standing waves. PMC walls are transparent to these radio waves and thus escape through these walls.

2.7.3 HELMHOLTZ EQUATION SOLUTION

$\nabla^2\Psi + k^2\Psi = 0$ (source less medium) where Ψ is a wave function and k is a wave number.

$$\nabla^2\Psi + k^2\Psi = -\mu J \text{ (medium with source)}$$

Maxwell's equations:

$$\nabla \times E = -\mu\frac{\partial H}{\partial t} \quad \nabla \times H = J + \epsilon\frac{\partial E}{\partial t}$$

Solving L.H.S. of both sides first:

$$\nabla \times E = \begin{vmatrix} i & j & k \\ \dfrac{\partial}{\partial x} & \dfrac{\partial}{\partial y} & \dfrac{\partial}{\partial z} \\ E_x & E_y & E_z \end{vmatrix} = i\left(\frac{\partial Ez}{\partial y} - \frac{\partial Ey}{\partial z}\right) - j\left(\frac{\partial Ez}{\partial x} - \frac{\partial Ex}{\partial z}\right) + k\left(\frac{\partial Ey}{\partial x} - \frac{\partial Ex}{\partial y}\right)$$

$$\nabla \times H = \begin{vmatrix} i & j & k \\ \dfrac{\partial}{\partial x} & \dfrac{\partial}{\partial y} & \dfrac{\partial}{\partial z} \\ H_x & H_y & H_z \end{vmatrix} = \left(\frac{\partial Hz}{\partial y} - \frac{\partial Hy}{\partial z}\right) - j\left(\frac{\partial Hz}{\partial x} - \frac{\partial Hx}{\partial z}\right) + k\left(\frac{\partial Hy}{\partial x} - \frac{\partial Hx}{\partial y}\right)$$

Comparing with R.H.S. in both equations and getting value of H_x, H_y, H_z from above and E_x, E_y, E_z from the above equations, we get:

$$Hx = \frac{1}{-j\omega\mu}\left(\frac{\partial Ez}{\partial y} - \frac{\partial Ey}{\partial z}\right) \quad Ex = \frac{1}{j\omega\epsilon}\left(\frac{\partial Hz}{\partial y} - \frac{\partial Hy}{\partial z}\right)$$

$$Hy = \frac{1}{j\omega\mu}\left(\frac{\partial Ez}{\partial x} - \frac{\partial Ex}{\partial z}\right) \quad Ey = \frac{1}{-j\omega\epsilon}\left(\frac{\partial Hz}{\partial x} - \frac{\partial Hx}{\partial z}\right)$$

$$Hz = \frac{1}{-j\omega\mu}\left(\frac{\partial Ey}{\partial x} - \frac{\partial Ex}{\partial y}\right) \quad Ez = \frac{1}{j\omega\epsilon}\left(\frac{\partial Hy}{\partial x} - \frac{\partial Hx}{\partial y}\right)$$

Substituting propagation coefficient $\dfrac{\partial}{\partial z} = \gamma$, we get:

$$Hx = \frac{j\omega\epsilon\dfrac{\partial Ez}{\partial y} + \gamma\dfrac{\partial Hz}{\partial x}}{\gamma^2 + \omega^2\mu\epsilon} \quad Ex = \frac{-j\omega\mu\dfrac{\partial Hz}{\partial y} + \gamma\dfrac{\partial Ez}{\partial x}}{\gamma^2 + \omega^2\mu\epsilon}$$

$$Hy = \frac{-j\omega \int \dfrac{\partial Ez}{\partial x} + \gamma \dfrac{\partial Hz}{\partial y}}{\gamma^2 + \omega^2 \mu\epsilon} \qquad Ey = \frac{j\omega\mu \dfrac{\partial Hz}{\partial x} + \gamma \dfrac{\partial Ez}{\partial y}}{\gamma^2 + \omega^2 \mu\epsilon}$$

Substituting in the above equation, we get:

$$-\left(\gamma^2 + \omega^2 \mu\epsilon\right) Hz = \frac{\delta^2 Hz}{\delta x^2} + \frac{\delta^2 Hz}{\delta y^2}$$

$$-\left(\gamma^2 + \omega^2 \mu\epsilon\right) Ez = \frac{\delta^2 Ez}{\delta x^2} + \frac{\delta^2 Ez}{\delta y^2}$$

$\nabla^2 \Psi + k^2 \Psi = 0$, Here, k is a wave number and $k^2 = k_x^2 + k_y^2 + k_z^2$.

$$\Psi = \Psi_x . \Psi_y . \Psi_z$$

$$\nabla^2 \Psi = \Psi\left(\frac{1}{\Psi_x}\left(\frac{\partial}{\partial x}\right)^2 \Psi + \frac{1}{\Psi_y}\left(\frac{\partial}{\partial y}\right)^2 \Psi + \frac{1}{\Psi_z}\left(\frac{\partial}{\partial z}\right)^2 \Psi\right)$$

Substituting the values of $\nabla^2 \Psi$ and k^2 from above, we get:

$$\Psi\left(\frac{1}{\Psi_x}\left(\frac{\partial}{\partial x}\right)^2 \Psi + \frac{1}{\Psi_y}\left(\frac{\partial}{\partial y}\right)^2 \Psi + \frac{1}{\Psi_z}\left(\frac{\partial}{\partial z}\right)^2 \Psi\right) + \left(k_x^2 + k_y^2 + k_z^2\right)\Psi = 0$$

Separating the independent terms, we get:

$$\frac{1}{\Psi_x}\left(\frac{\partial}{\partial x}\right)^2 \Psi = -k_x^2$$

$$\frac{1}{\Psi_y}\left(\frac{\partial}{\partial y}\right)^2 \Psi = -k_y^2$$

$$\frac{1}{\Psi_z}\left(\frac{\partial}{\partial z}\right)^2 \Psi = -k_z^2$$

$$\Psi \text{ or } Hz \text{ or } Ez = \left\{\left(A\sin k_x.x + B\cos k_x x\right)\left(C\sin k_y.y + D\cos k_y.y\right)\right\}e^{-jk_z z}$$

$$Hz = \sum_{m,n}\left\{C_{mn}\left(\cos\frac{m\pi x}{a}\right)\left(\cos\frac{n\pi y}{b}\right)\right\}e^{-jk_z z} \; ; \; C_{mn} \text{ Fourier coefficients}$$

$$Ez = \sum_{m,n} \left\{ D_{mn} \left(\sin \frac{m\pi x}{a} \right) \left(\sin \frac{n\pi y}{b} \right) \right\} e^{-jk_z z}; \ D_{mn} \text{ Fourier coefficients}$$

Hence, Ez and Hz fields are solutions of DRA as propagating fields (Figures 2.16 through 2.18).

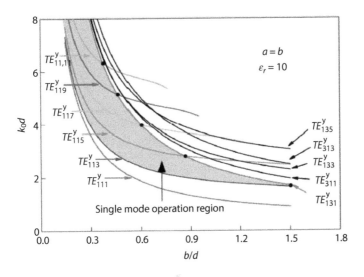

FIGURE 2.16   Higher order modes and aspect ratio relationship.

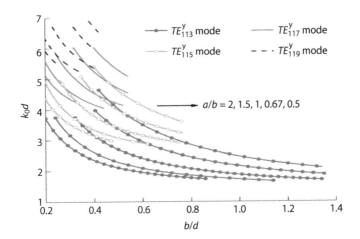

FIGURE 2.17   Higher order modes and aspect ratio (a/b fixed).

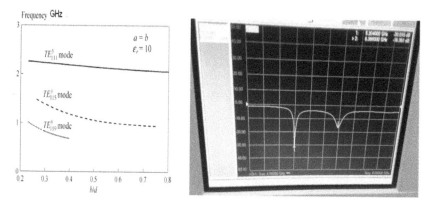

FIGURE 2.18   Higher order mode.

2.8   BEAM WIDTH CONTROL IN DRA

Axial ratio control using an absorber can help to control the beam width of an antenna as aspect ratio gives rise to higher mode generations. Operating at fundamental mode shall provide a particular beam width. This can be seen in 2D radiation patterns. When the same DRA is operated with higher order modes, the beam width shall get reduced and gain will be enhanced. This way DRA beamwidth control can be achieved. One can say that directivity is more when beamwidth is less and vice versa.

Radiation power was regulated in a particular direction or angle using multi-layer high and low permittivity materials. High permittivity material has high loss tangent and low permittivity material has low loss tangent. These are also known as meta materials. Absorbers were used to model for beam width control. Given below are DRA model results obtained based on above stated facts. Use of concave type superstrate, when placed on top of DRA with optimum air gap, has controlled the beamwidth of DRA (Figures 2.19 through 2.22).

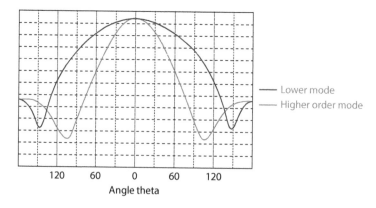

FIGURE 2.19   Generation of beamwidth control with concave superstrate placed on top of DRA.

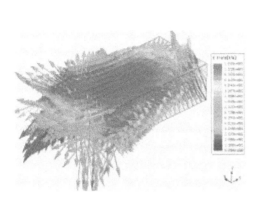

FIGURE 2.20 E-field generation and antenna under measurements in anechoic chamber.

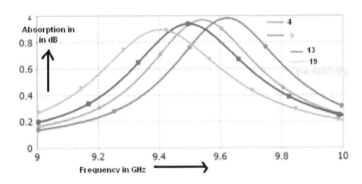

FIGURE 2.21 Showing absorption loss, frequency shift due to change in permittivity of DRA.

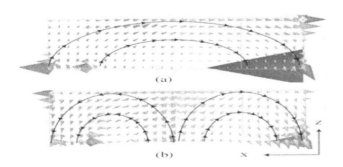

FIGURE 2.22 Fundamental and higher order modes generation in DRA (a) fundamental mode and (b) higher order mode.

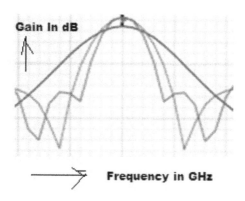

FIGURE 2.23 Sensitivity analysis and beamwidth control in RDRA.

Shift in frequency at different permittivity frequency is inversely proportional to permittivity Ɛr and power absorption is high when permittivity is high. Beam width at higher order mode in rectangular DRA is narrow and broad at lower order mode. Beam width is narrow at higher order operating mode (Figure 2.23).

This can be achieved using fluid frame DRA; as height will increase, higher order mode will generate and beam width will be narrow. Thus beam width control can be achieved. Here beam width is controlled using concave superstrates on top of DRA. Beam width can be controlled by antenna placed in arrays, frequency selective surface (FSS) designs, and multi input multi output (MIMO) designs (Figures 2.24 through 2.28).

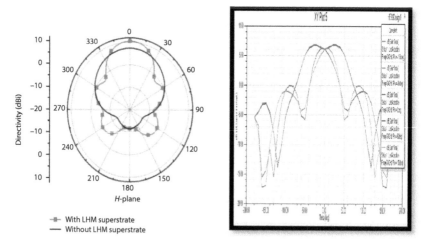

FIGURE 2.24 Superstrate layer as possible use to control beam width of DRA.

FIGURE 2.25 Defected ground plane structure (DGS) as dumbbell shape can be used to reduce resonant frequency of DRA.

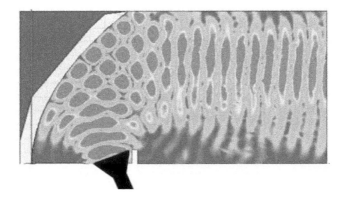

FIGURE 2.26 Feed horn antenna showing reflected mode of propagation can be used to focus beam of an antenna.

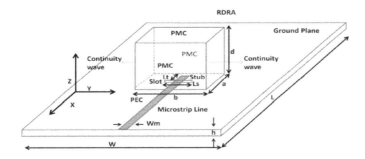

FIGURE 2.27 Aperture coupled rectangular DRA.

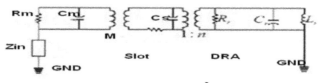

EQUIVALENT CIRCUIT of ANTENNA

FIGURE 2.28 Equivalent circuit of aperture coupled DRA.

Dielectric wave model gives a solution for rectangular DRA with a characteristic equation and a transcendental equation as given below:

Characteristic equation:

$$\epsilon_r\, ko^2 = kx^2 + ky^2 + kz^2 \; ; k = \omega\sqrt{\mu\epsilon}$$

where $kx = \frac{m\pi}{a}$; $ky = \frac{n\pi}{b}$

Transcendental equation:

$$kz\left(\tan k_z \frac{d}{2}\right) = \sqrt{(\epsilon_r - 1)k_o^2 - k_z^2}$$

The coupling of DRA depends on the position of the DRA placed with reference to slot:

$$fr = \frac{3\times10^8 \times 10^3}{2\times\sqrt{12.8}} \sqrt{\left(\frac{1}{12}\right)^2 + \left(\frac{1}{12}\right)^2 + \left(\frac{1}{2\times15}\right)^2} = 4.4\ \text{GHz}$$

If $a = b = 4.54$
$d = 4.68$
$\epsilon_r = 12.8$

$$fr_{111} = 9.07\ \text{GHz}$$

$$\frac{f_r}{f_2 - f_1} = 2\pi \frac{We}{\text{prad}} = Q = \frac{\omega o We}{\text{Prad}} = 2\pi\left(\frac{\text{Time energy stores}}{\text{Energy loss per cycle}}\right)$$

$$\frac{\text{Prad}}{|I|^2} = Z = \frac{V_{in}}{I_{in}}; \ \text{VSWR} = \frac{V_{max}}{V_{min}} = Z$$

$$Z_o = R + JX$$

$$\sqrt{Z_{sc}\,Z_{oc}} = Z_o = \sqrt{\frac{L}{C}}; \quad \delta\,(\text{skin depth}) = \frac{1}{\sqrt{\pi f \mu \sigma}}$$

$$Z = \frac{P}{|I|^2}$$

$$\text{Prad} = \frac{|E_\theta|^2 + |E_\phi|^2}{\eta}$$

2.8.1 MICROSTRIP LINE

The dimensions for W-width and L-length of microstrip are given as follows:

$$W = \frac{c}{2 f_r \sqrt{\dfrac{\epsilon_r + 1}{2}}}$$

Width of feed line, $W = \dfrac{c}{2 f_r \sqrt{\dfrac{\epsilon_r + 1}{2}}}$

$$\epsilon_{ff} = \frac{\epsilon_r + 1}{2} + \frac{\epsilon_r - 1}{2}\left(\frac{1}{\sqrt{1 + 12\left(\dfrac{h}{w}\right)}}\right)$$

$$\Delta L = 0.412\,h\left(\frac{\epsilon_{eff} + 0.3}{\epsilon_{eff} - 0.258}\right)\left(\frac{\dfrac{W}{h} + 0.262}{\dfrac{W}{h} + 0.813}\right)$$

$$L_{eff} = \frac{c}{2 f_r \sqrt{\epsilon\,r_{eff}}}$$

Quality factor of cylindrical DRA is dependent on aspect ratio (radius/height) and permittivity (Figures 2.29 and 2.30).

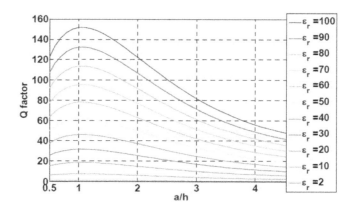

FIGURE 2.29 Quality factor increases at high permittivity.

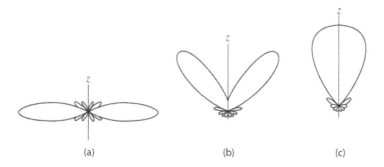

FIGURE 2.30 Radiation patterns of DRA. (a) Broadside, (b) intermediate, and (c) end fire.

2.9 CYLINDRICAL DRA RESONANT FREQUENCY

2.9.1 Resonant Frequency

$$\left(\frac{F_{TMnpm}}{F_{TMnpm}}\right) = \frac{c}{2\pi a\sqrt{\varepsilon_r\,\mu_r}}\sqrt{\left(\frac{X_{np}}{X'_{np}}\right)^2 + \left(\frac{(2m+1)\pi.a}{2d}\right)^2}$$

where X_{np} and X'_{np}, are Bessel's solutions and $(n, m, p) \in N^3$ and a, d are the radius and the height of the cylindrical dielectric resonator antenna. $X'_{11} = 1.841$ for fundamental mode (Figure 2.31). $X'_{11} = 1.841$ for fundamental mode (Figure 2.31).

The fundamental excited mode is the $HE_{11\delta}$ and its resonant frequency equals

$$f_{110} = \frac{3.10^8}{2\pi\sqrt{30}}\sqrt{\left(\frac{X'_{11}}{0.04}\right)^2 + \left(\frac{\pi}{2\times0.045}\right)^2} = 503.6\,\text{MHz}$$

The orientation of DRA and excitations has an impact on resonant modes (Figures 2.32 and 2.33).

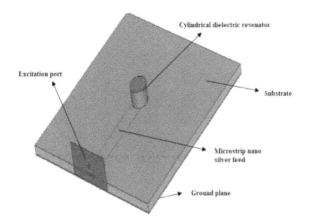

FIGURE 2.31   Cylindrical DRA.

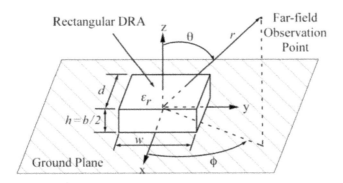

FIGURE 2.32   Rectangular DRA showing far fields.

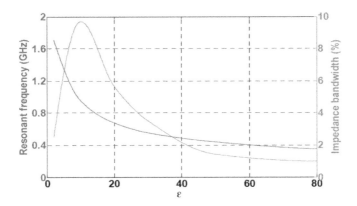

FIGURE 2.33   Bandwidth and permittivity relationship.

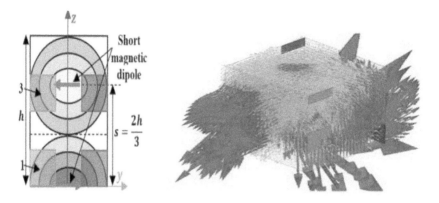

FIGURE 2.34 Rectangular DRA short magnetic dipole, E-field distribution and prototype designed to operate at 6.5 GHz.

Rectangular DRA fields coming out from transparent walls (PMC). The resonant modes are excited into DRA. The Number of short magnetic dipoles inside DRA presents excited modes (Figure 2.34).

2.9.2 DESIGN DIMENSIONS

$wg = lg = 50$, $a = b = h = 12$, $lm = 25$, $s = 5$, $wm = 2.2$, $ls = 9$, $ws = 1,1 = w2 = 3.6$, h1 $= 3$, h2 $= 1.3$ (all units in mm, Face: 1-front, 2-right, 3-back and 4-left, copper strips at front face and blue strips at back face, $L1$–$L4$ are layers 1 to 4).

Substrate length $Lg = L + 6h$; where h is substrate width $(Wg) = W + 6h + 6h$ $Ng = b + 4h$; due to fringing effect the width of the ground plane is larger than RDRA. The resonant frequency with eigen states of DRA can be expressed for isolated "$2d$" will be replaced by "d."

2.10 RECTANGULAR DRA THEORY

Resonant modes are spectral resolution of electromagnetic fields of waves radiated by RDRA. These modes are field structures that can exist inside DRA. Modes are patterns of electromagnetic fields which repeat themselves in a sinusoidal pattern. They are dependent upon the following parameters:

1. Input excitation.
2. Mediums or dielectric used.
3. Coupling method used.
4. Point of excitation in DRA.
5. Input impedance of DRA.
6. Can be analyzed by short magnetic dipoles.
7. DRA radiates like horizontal magnetic dipole.
8. Resonant modes are used for prediction of radiation pattern.
9. Superstrates are used to control directivity, gain and beam width control.

10. Active devices in superstrates can be integrated for beam width control.
11. Use of concave and convex superstrate is an innovative approach for beam width control in an antenna.
12. Use of multiple permittivity materials for proper control of beam.
13. Plasmonic DRA with Gaussian beam with silver nano waveguide.
14. Possibility of excitation of fundamental, higher order and hybrid modes.
15. Modes can be controlled and manipulated by proper choice of DRA dimensions.
16. Merging for wide band, separation for multiband and higher order mode for high gain and DE miniaturization of DRA.
17. Beam width control applications can be made at airport advance landing system and vehicular communication to combat bad weather.
18. TE_{mnp}^{y} can be excited for odd modes only aspect ratio b/d govern excitation of resonant modes field pattern formed inside DRA is resonant mode and same will be reflected in far field pattern of an antenna.

2.11   MINIATURIZATION IN RECTANGULAR DRA DUE TO METALLIC STRIPS

The miniaturized of DRA is obtained by using the metallic strips on the side walls of the DRA. These metallic strips excite the orthogonal degenerate modes. The metallic strips are on four different heights, $L1$ to $L4$, on the different faces of the DRA. The DRA is excited for first and third order modes. The objective is to rotate the field vectors in plane. Surface area of the metallic strips decides field vectors. In fundamental mode, the field is confined near to the slot in rectangular DRA. The frequency ratio of DRA can also be tuned. This is achieved by changing the dimensions of metallic strips. The location and size of the applied metallic strips also matters (Figure 2.35).

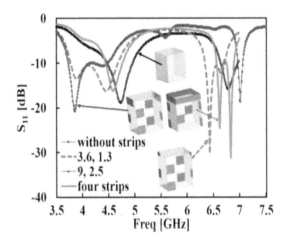

FIGURE 2.35   Rectangular DRA with metal strips.

2.11.1 DIELECTRIC BOUNDARY CONDITIONS IN DRA

The propagating field will escape through transparent walls only and will get reflected with metallic walls. This reflection and refraction forms field patterns as resonant modes inside the DRA. The tangential E component is continuous across interfaces, both dielectric-to-dielectric and PEC-to-dielectric. The normal H component is continuous across dielectric interfaces; it is discontinuous across PEC-to-dielectric interfaces due to the presence of free surface charge PEC and PMC boundaries.

Because the coupling of the mode depends on both length and height of the probe, tangential and normal components of E and h fields are:

The boundary conditions are expressed as $\begin{aligned} n \times E &= 0 \\ n \cdot H &= 0 \end{aligned}$ for PEC.

And $\begin{aligned} n \times H &= 0 \\ n \cdot E &= 0 \end{aligned}$ for PMC.

Also as per law of conservation of energy: $\int_V |E|^2 \, dv = \int_V |H|^2 \, dv$ satisfy.

2.12 CIRCULAR POLARIZATION IN DRA

The E field rotates due to design curves introduced in DRA. This rotation can be adjusted into RHCP and LHCP patterns. These pattern can be named as right hand circular polarization or left hand circular polarization. The orientation of electric fields with respect to propagation of electromagnetic waves is called polarization. This can be vertical, horizontal, elliptical or circular polarization. A circular polarized signal is called a robust signal in propagation. This chapter describes a single micro-strip fed circularly polarized dielectric resonator antenna. The dielectric resonator antenna is given circular polarization by introducing electric current to cause rotation.

The superposition of two plane waves are:

$$\vec{E} = (\hat{x}E_1 + \hat{y}E_2)e^{-jkz}$$

For circular polarization:
   (E_1 and E_2 are orthogonal)

$$\rightarrow E_1 = jE_2 = E_o$$

$$\rightarrow \vec{E} = E_o\left(\hat{x} - j\hat{y}\right)e^{-jkz}$$

$$\vec{E} = \left[\hat{x}\cos\left(\omega t - kz\right) - \hat{y}\cos\left(\omega t - kz - \frac{\pi}{2}\right)\right]$$

At $z = 0$, $\vec{E} = \left[\hat{x}\cos\left(\omega t\right) - \hat{y}\cos\left(\omega t - \frac{\pi}{2}\right)\right]$

The variation in t from 0 to T will cause rotation of E field vectors from 0 to 2π (Figures 2.36 and 2.37).

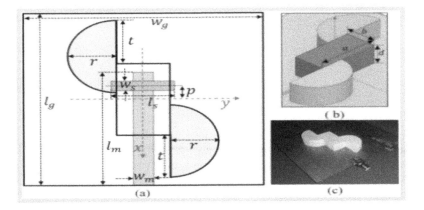

FIGURE 2.36 (a) Circular polarization implementation architecture of antenna, (b) HFSS model of antenna, and (c) hardware model of antenna.

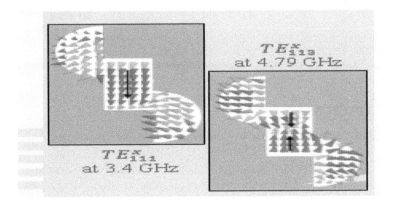

FIGURE 2.37 E-fields showing circular polarization.

2.12.1 LHCP/RHCP

Field rotations are achieved by shape of slot. Again it can be RHCP or LHCP depending on the dominant side of slot turn extra. The dominant arm of z shape will rotate other arm fields in the dominant direction (Figure 2.38).

2.12.2 DRA Fabrication Stages

DRAs, fabrication and development process is shown in Figure 2.39.

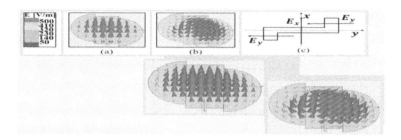

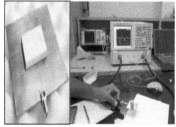

FIGURE 2.38   Circular polarization in DRA due to slot shape. (a) E-field with rectangular slot, (b) E-field with stair-shaped slot, and (c) mechanism of CP.

FIGURE 2.39   DRAs, fabrication stage to development, and then test and measurements.

2.13   CONCLUSION

DRA design is simple. It can excite multimodes and it has low loss. It is an efficient radiator. Its design has good flexibility due to different aspect ratios available to the designer. It possess low loss due to use of dielectric material as compared to metal. High gain along large band width becomes possible in DRA. Merging of modes provides more bandwidth in DRA. Use of concave type superstrate in DRA provides a narrow beam. Control on modes has given reconfigurable parameters in DRA. Large bandwidth, high gain and narrow beam width are excellent features of DRA. Hence, the concept of versatile antenna designs has become possible to antenna designers, with the advent of DRA. Designing nano DRAs can further boost these advantages for biomedical imaging and airport security applications. Modes in DRA can be controlled for high gain, wide bandwidth, less beam width and multiple modes with LHCP and RHCP options. The size of DRA is miniaturized using metallic strips.

REFERENCES

1. M. Ansarizadeh and A. Ghorbani, "An Approach to Equivalent Circuit Modeling of Rectangular Microstrip Antennas," *Prog. Electromag. Res. B*, vol. 8, pp. 77–86, 2008.
2. R. M. van Schelven., "Equivalent Circuit Models of Finite Slot Antennas," *IEEE Trans. Antennas Propag.* doi:10.1109/TAP.2019.2908112.

3. S. Sun, T. S. Rappaport, M. Shafi, P. Tang, J. Zhang, and P. J. Smith., "Propagation Models and Performance Evaluation for 5G Millimeter-Wave Bands," *IEEE Trans. Vehicular Technol.*, 2018.

4. J. Medbo et al., "Radio Propagation Modeling for 5G Mobile and Wireless Communications," *IEEE Commun. Mag.*, vol. 54, no. 6, pp. 144–151, 2016.

5. S. Hur et al., "Millimeter Wave Beamforming for Wireless Backhaul and Access in Small Cell Networks," *IEEE Trans. Commun.*, vol. 61, no. 10, pp. 4391–4403, 2013.

6. C. E. Zebiri, M. Lashab, D. Sayad, I. T. E. Elfergani, K. H. Sayidmarie, F. Benabdelaziz, R. A. Abd-Alhameed, J. Rodriguez, J. M. Noras, "Offset Aperture-Coupled Double-Cylinder Dielectric Resonator Antenna With Extended Wideband," *IEEE Trans. Antennas Propag.*, 2017.

7. R. S. Yaduvanshi and H. Parthasarathy, *Rectangular Dielectric Resonator Antennas Theory and Design*. New Delhi: Springer, 2016.

8. G. Varshney, V. S. Pandey, R. S. Yaduvanshi, L. Kumar, "Wide Band Circularly Polarized Dielectric Resonator Antenna with Stair-Shaped Slot Excitation," *IEEE Trans. Antennas Propag.*, 2016.

9. M. Singh, R. S. Yaduvanshi, A. Vaish, "Design for Enhancing Gain in Multimodal Cylindrical Dielectric Resonator Antenna," *IEEE Conference.*, December 2015.

10. S. Gotra, G. Varshney, R. S. Yaduvanshi, V. S. Pandey, "Dual-Band Circular Polarisation Generation Technique with the Miniaturization of a Rectangular Dielectric Resonator Antenna," *IET Microwaves, Antennas Propag.*, 2019.

11. M. Singh, A. K. Gautam, R. S. Yaduvanshi, A. Vaish, "An Investigation of Resonant Modes in Rectangular Dielectric Resonator Antenna Using Transcendental Equation," *Wireless Pers. Commun.*, vol. 95, no. 3, pp. 2549–2559, 2017.

12. G. Bakshi, A. Vaish, R. S. Yaduvanshi, "Two-Layer Sapphire Rectangular Dielectric Resonator Antenna for Rugged Communications," *Prog. Electromagn. Res. Lett.*, vol. 85, pp. 73–80, 2019.

13. G. Kumar, M. Singh, S. Ahlawat, R. S. Yaduvanshi, "Design of Stacked Rectangular Dielectric Resonator Antenna for Wideband Applications," *Wireless Pers. Commun.*, vol. 109, no. 3, pp. 1661–1672, 2019.

14. R. Khan, "Multiband-Dielectric Resonator Antenna for LTE Application," *IET Microwaves Antennas Propag.*, 2016.

15. R. D. Richtmyer, "Dielectric Resonators," *Journal of Applied Physics*, vol. 10, no. 6, pp. 391–398, 1939.

3 Resonant Frequency Computations of DRA

3.1 INTRODUCTION

Dielectric resonator antennas (DRA) have advantages, such as low cost, high radiation efficiency, large operating bandwidth, low profile, and low losses. They are compatible to generate higher order modes. The dielectric waveguide model is generally used to compute resonant frequency of DRA [1–13]. The waveguides or strip lines along conducting walls of DRA disturb the external fields, and mathematical equations are developed to calculate the resonant frequency accurately. Guillon et al. developed a new method based upon a mixture of the magnetic walls, and the dielectric waveguide models calculated resonant frequency of rectangular DRA. Legier et al. used Marcatili's model and Knox and Toulios's model in dielectric waveguide model to calculate the resonant frequency of rectangular DRA. Mongia et al. introduced a conventional wave guide model (CWGM). The resonant frequency of rectangular DRA is a function of two aspect ratios and permittivity. The concept of image theory was applied and replaced the resonator with an isolated dielectric resonator.

Antar introduced a modified wave guide model (MWGM), which expressed the resonant frequency in a function of effective dimensions of rectangular DRA. Based on the concept of perturbation theory, a modification to the dielectric waveguide model (DWM) was presented in 2016 for the calculation of resonant frequency of rectangular dielectric resonator antenna (DRA). Furthermore, there is a wide difference between theoretical and experimental resonant frequencies calculated from these models.

In this chapter, the concept of effective dimension of rectangular DRA is introduced which provides more accurate calculation for resonant frequency of rectangular DRA. The physical dimension of DRA is defined as the dimension of DRA when no voltage-current supply is applied at the input connector. When voltage-current supply is applied at the input connector, then due to electric field, fringing effects appeared in DRA, resulting in the effective dimension of DRA being increased, and this new dimension of DRA is known as the effective dimension. Transcendental equation is used for mathematical computation of resonant frequency, which is similar to experiment results. The relation between effective dimensions and their physical dimensions is introduced. The measured results of resonant frequency of rectangular DRA are compared with calculated results using a transcendental equation.

3.2 MATHEMATICAL MODELING

Figure 3.1 shows the geometry of a rectangular DRA. The rectangular DRA has three dimensions (length, width, and height), that is, two aspect ratio, providing more degrees of freedom than for the hemispherical and cylindrical counterparts. In this paper, the slot-coupled rectangular DRA is investigated.

The rectangular DRA of length a, width b, and height d is designed. Aperture coupling is used to excite the antenna where the slot is along the y direction.

The transcendental equation is derived below. It is a non-confined mode and does not satisfy $\hat{n} \times \vec{H} = 0$ on all surfaces. Hence, calculation of the resonant frequency is based on a waveguide structure with a perfect electric conductor (PEC) plane at $z = 0$ and perfect magnetic conductor (PMC) at the $x = \pm a/2$ and $z = d$.

The fields analysis is given below:

$$E_x = \frac{1}{j\omega\varepsilon\left(1+\frac{\gamma^2}{k^2}\right)}\left[\frac{\partial H_z}{\partial y} - \frac{1}{j\omega\mu}\frac{\partial^2 E_z}{\partial z \partial x}\right] \tag{3.1}$$

$$E_y = \frac{1}{j\omega\varepsilon\left(1+\frac{\gamma^2}{k^2}\right)}\left[-\frac{1}{j\omega\mu}\frac{\partial^2 E_z}{\partial z \partial y} - \frac{\partial H_z}{\partial x}\right] \tag{3.2}$$

$$H_x = \frac{-1}{j\omega\mu\left(1+\frac{\gamma^2}{k^2}\right)}\left[\frac{\partial E_z}{\partial y} - \frac{1}{j\omega\varepsilon}\frac{\partial^2 H_z}{\partial z \partial x}\right] \tag{3.3}$$

$$H_y = \frac{-1}{j\omega\mu\left(1+\frac{\gamma^2}{k^2}\right)}\left[\frac{1}{j\omega\varepsilon}\frac{\partial^2 H_z}{\partial z \partial y} - \frac{\partial E_z}{\partial x}\right] \tag{3.4}$$

where:

$E_{x,\,y}$ and $H_{x,\,y}$ are the electric and magnetic field in x and y directions
ω is the angular frequency
μ is the permeability
ε is the permittivity
γ is an arbitrary constant
k is the function of wave number

Now the propagating wave is continuous at the interface, that is

$$H_y = H'_y$$

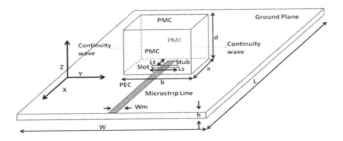

FIGURE 3.1 Geometries of rectangular DRA (RDRA).

Therefore,

$$A\cos\left(\frac{m\pi}{a}x\right)\sin\left(\frac{n\pi}{b}y\right)\left(C_1 e^{jk_z z}+C_2 e^{-jk_z z}\right)$$

$$=A\cos\left(\frac{m\pi}{a}x\right)\sin\left(\frac{n\pi}{b}y\right)C'_2 e^{-jk'_z Z}\text{ at }z=0,$$

or $C_1 e^{jk_z Z}+C_2 e^{-jk_z Z}=C'_2 e^{-jk'_z Z}$ at $z=0$ (3.5)

But at $z = 0$, only the inside waveform exists.
  Therefore,

$$C_1 e^{jk_z z}+C_2 e^{-jk_z z}=0\text{ at }z=0$$

Now substituting the value of $z = 0$, we get

$$C_1 + C_2 = 0\text{ or }C_1 = -C_2$$

As H_z is a continuous magnetic field about the interface $z = d$ and H'_z is a magnetic field outside the DRA, i.e., for $z > d$, therefore

$$H_z = H'_z\text{ at }z = d,$$

and

$$\frac{\partial H_z}{\partial z}=\frac{\partial H'_z}{\partial z}\text{ at }z = d, (\text{as }H_x = H'_x\text{ at }z = d)$$

Since $C_1 = -C_2$, we can write

$$H_z = B\sin\left(\frac{m\pi}{a}x\right)\sin\left(\frac{n\pi}{b}y\right)\sin(k_z z)$$

and

$$H'_z = B \sin\left(\frac{m\pi}{a} x\right) \sin\left(\frac{n\pi}{b} y\right) \sin(k'_z z)$$

So,

$$B \sin\left(\frac{m\pi}{a} x\right) \sin\left(\frac{n\pi}{b} y\right)\left(C_1 e^{jk_z z} - C_2 e^{-jk_z z}\right) = B \sin\left(\frac{m\pi}{a} x\right) \sin\left(\frac{n\pi}{b} y\right) C'_2 e^{-jk'_z Z}$$

or

$$C_1 e^{jk_z z} - C_2 e^{-jk_z z} = C'_2 e^{-jk'_z Z}$$

From the equation, i.e.,

$$C_1 = -C_2 \tag{3.6}$$

at $z = d$,

$$2C_1 \cos(k_z d) = C'_2 e^{-jk_z d}$$

On equating $H_z = H'_z$

$$jk_z \left(C_1 e^{jk_z z} + C_2 e^{-jk_z z}\right) = -jk'_z C'_2 e^{-jk'_z Z}$$

$$2 jk_z C_1 \sin(k_z d) = -k'_z C'_2 e^{-jk'_z Z} \tag{3.7}$$

$$jk_z \tan k_z d = -k'_z$$

Squaring both sides and substituting the value of k'^2_z,

$$k'^2_z = k^2_z - \omega^2 \mu(\epsilon_r - 1)$$

and substituting $\mu = 1$, we get, in the case of isolated DRA,

$$k_z \tan(k_z d) = \sqrt{(\epsilon_r - 1)k_0^2 - k_z^2} \tag{3.8}$$

It is assumed that rectangular DRA is fed in this manner that the y direction has minimum dimension. The following transcendental equation is derived from (3.4):

$$k_y \tan\left(\frac{k_y d}{2}\right) = \sqrt{(\epsilon_r - 1)k_0^2 - k_y^2} \tag{3.9}$$

However, (3.5) can also be written as follows:

$$k_y b + 2 \tan^{-1}\left(\frac{k_y}{k_{y0}}\right) = n\pi \tag{3.10}$$

where $k_{y0} = \sqrt{(\epsilon_r - 1)k_0^2 - k_y^2}$

By binomial expansion of (3.6), it can be obtained:

$$k_y b + 2 k_y \left(\frac{\sqrt{(\epsilon_r - 1)k_0^2 - k_y^2}}{(\epsilon_r - 1)k_0^2}\right) = n\pi \tag{3.11}$$

The following equation is obtained by rearranging the variables:

$$k_y = \frac{n\pi}{b\left[1 + 2\left(\dfrac{1}{bk_0\sqrt{(\epsilon_r - 1)}}\right)\right]} \tag{3.12}$$

$k_y = {}^{n\pi}\!/_{\hat{b}}$, where $\hat{b}$ is the effective width.

The relation between b (physical width) and $\hat{b}$ (effective width) can be expressed as follows:

$$\hat{b} = b + \left(\frac{2}{k_0\sqrt{(\epsilon_r - 1)}}\right) \tag{3.13}$$

The effective width is always greater than physical width, and the amount of increment in effective width depends on the dielectric constant value.

For TE$_{mnp}^y$ mode, the m, n, and p are integer numbers of a half wave variation in x, y, and z directions, respectively. For the case of TE$_{1\delta1}^y$ mode, when the half wave variation in the y direction, i.e., n is not integer and is less than 1, then a new parameter δ is introduced. δ is a fraction of a half wave variation in the y direction, the value of which varies from 0 to 1.

$$\delta = \frac{k_y}{(\pi/b)} \tag{3.14}$$

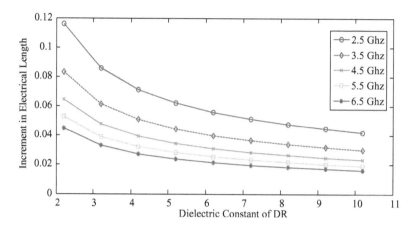

FIGURE 3.2   Effective width of DRA with frequency and dielectric constant.

From (3.8) and (3.10), δ can be derived as follows (Figure 3.2):

$$\delta = \cfrac{1}{\left(1+2\left(\cfrac{1}{bk_0\sqrt{(\epsilon_r - 1)}}\right)\right)} \qquad (3.15)$$

The orientation of rectangular DRA also affects the resonant frequency.

3.3   MEASURED AND SIMULATED RESULTS

DRA size of 30×20 mm^2 and a relative permittivity constant of 2.2 is used in experimentation (Table 3.1).

TABLE 3.1
Dimensions of DRA

S. No.	Name of Element	Dimension (mm)
1.	Ground plane ($L \times W$)	30×20
2.	Substrate height (h)	0.8
3.	DR dimension ($a \times b \times d$)	$9 \times 4.6 \times 10.8$
4.	Length of slot (L_s)	3.743
5.	Width of slot (W_s)	0.404
6.	Length of strip (L_m)	13.356
7.	Length of stub (L_t)	5.337

3.3.1 EFFECT OF OPERATING FREQUENCY (MODES) OF RECTANGULAR DRA

When input is applied, then three modes in rectangular DRA are observed at resonant frequency of 8.4422, 12.84, and 16.67 GHz as shown in Figure 3.3.

It is observed that at lower resonant frequencies, as shown in Figure 3.4a (f_o = 8.4433 GHz), the electric fields are not confined to the center of rectangular DRA. As resonant frequency is increased, as shown in Figure 3.4b (f_o = 12.7763 GHz) and Figure 3.4c (f_o = 16.6708 GHz), the electric field is confined at the center of rectangular DRA. It has also been observed that effective dimension (width) decreases with high resonant frequencies or modes.

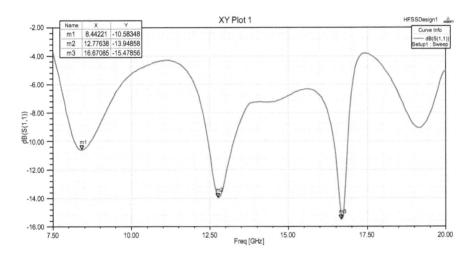

FIGURE 3.3 Return loss of rectangular DR.

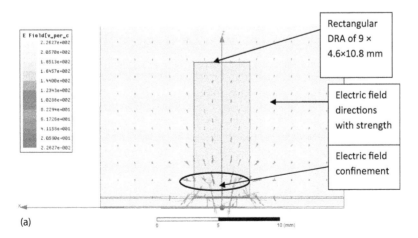

FIGURE 3.4 Simulated electric field distribution inside rectangular DRA with different values of operating frequency 8.4433 GHz at (a). *(Continued)*

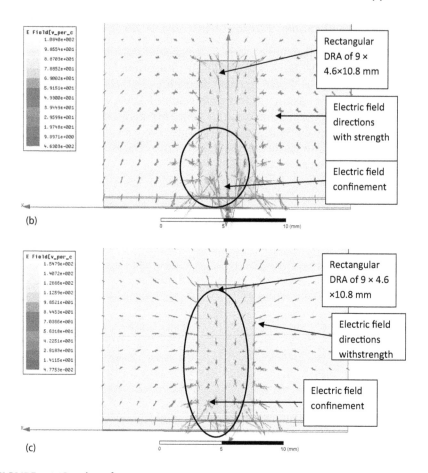

FIGURE 3.4 (Continued)   Simulated electric field distribution inside rectangular DRA with different values of operating frequency 12.7763 GHz at (b) and 16.6708 GHz at (c).

3.3.2   Effect of Dielectric Constant of Rectangular DRA Material

Effect on E-field distribution when DRA permittivity changes, is shown in Figure 3.5.

3.3.3   Effect of Orientation of Rectangular DRA

The rectangular DRA of given dimensions (9 × 4.6 × 10.8 mm) is placed on slot in all possible 06 orientations to get the results. These results are tabulated in Table 3.2.

Hence, orientation of rectangular DRA also affects the effective dimension of rectangular DRA.

Figure 3.6 is an experimental setup to calculate the resonant frequency of rectangular DRA (see also Figure 3.7).

The result shows the resonant frequency rectangular DRA of 14 × 15 × 16 mm is 4.7 GHz as shown in Figure 3.8.

Resonant frequency calculated from theoretical and simulated through HFSS is compared as shown in Figure 3.9.

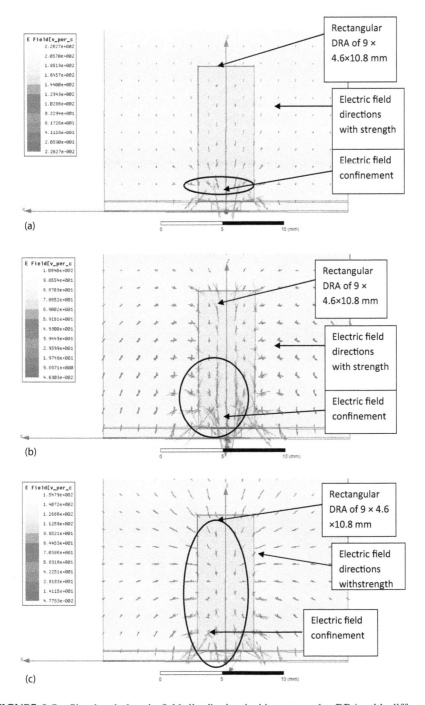

FIGURE 3.5 Simulated electric field distribution inside rectangular DRA with different values of dielectric constant 4.9 at (a), 9.8 at (b), and 14.7 at (c).

TABLE 3.2

Effect on Electric Length with DR Orientation in Rectangular DRA

S. No.	Orientation ($a \times b \times d$), mm	Calculated Resonant Frequency (Using Characteristic Equation), GHz	Calculated Resonant Frequency (Using Transcendental Equation), GHz	Measured Resonant Frequency Results, GHz
1.	$9 \times 4.6 \times 10.8$	11.907	7.771	7.72
2.	$4.6 \times 9 \times 10.8$	11.907	11.386	11.42
3.	$4.6 \times 10.8 \times 9$	11.631	11.306	11.31
4.	$10.8 \times 4.6 \times 9$	11.631	7.178	7.04
5.	$9 \times 10.8 \times 4.6$	8.669	8.113	8.21
6.	$10.8 \times 9 \times 4.6$	8.669	7.725	7.01

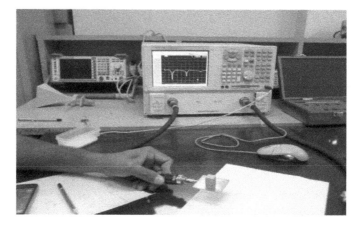

FIGURE 3.6 Photograph for frequency measurements of rectangular DRA.

FIGURE 3.7 Photograph for fabricated rectangular DRA.

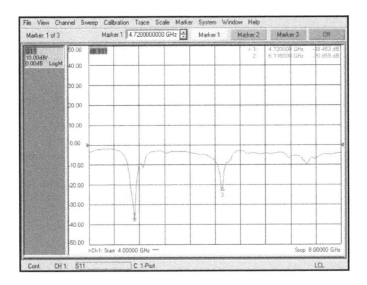

FIGURE 3.8 Snapshot for S_{11} of rectangular DRA during experimental setup.

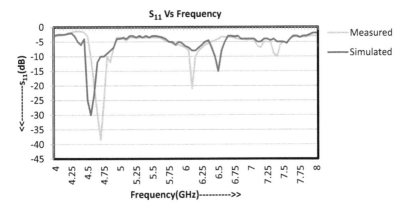

FIGURE 3.9 Comparison of simulation and measured results of rectangular DRA.

Hence, effective dimension provides more accurate results for calculation of resonant frequency in rectangular DRA.

3.4 CONCLUSION

Physical and effective dimensions of rectangular DRA for different modes have been
worked. Hence, electromagnetic energy confined within the rectangular DRA is
dependent on permittivity, resonance frequency, its physical dimensions, and orientation of rectangular DRA. The high value of relative permittivity of rectangular
DRA, E-fields, gets confined, so reducing effective dimension. In contrast, with low
permittivity use due to fringing and the effective distance between the E-field, effective dimension increases.

REFERENCES

1. A. Goldsmith, *Wireless Communications*. Cambridge, MA: Cambridge University Press, 2005.
2. D. M. Pozar, *Microwave Engineering*, 4th ed. New York: Johan Wiley & Sons, Inc., 2005.
3. R. S. Yaduvanshi and H. Parthasarathy, *Rectangular Dielectric Resonator Antenna Theory and Design*. New York: Springer, 2016.
4. A. Petosa and A. Ittipiboon, "Dielectric Resonator Antennas: A Historical Review and the Current State of the Art," *IEEE Antennas Propag. Mag.*, vol. 52, no. 5, pp. 91–116, 2010.
5. G. Varshney, R. S. Yaduvanshi, and V. S. Pandey, "Gain and Bandwidth Controlling of Dielectric Slab Rectangular Dielectric Resonator Antenna," *12th IEEE Int. Conf. Electron. Energy, Environ. Commun. Comput. Control (E3-C3), INDICON 2015*, vol. 3, pp. 6–9, 2016.
6. T. Chang and J. Kiang, "Bandwidth Broadening of Dielectric Resonator Antenna by Merging Adjacent Bands," *IEEE Trans. Antennas Propag.*, vol. 57, no. 10, pp. 3316–3320, 2009.
7. G. Varshney, P. Mittal, V. S. Pandey, R. S. Yaduvanshi, and S. Pundir, "Enhanced Bandwidth High Gain Micro-strip Patch Feed Dielectric Resonator Antenna," *Int. Conf. Comput. Commun. Autom.*, vol. 1, pp. 92–95, 2016.
8. R. K. Mongia and P. Bhartia, "Dielectric Resonator Antennas—A Review and General Design Relations for Resonant Frequency and Bandwidth," *Int. J. Microw. Millimeter-Wave Comput. Eng.*, vol. 4, no. 3, pp. 230–247, 1994.
9. G. Varshney, V. S. Pandey, R. S. Yaduvanshi, and L. Kumar, "Wide Band Circularly Polarized Dielectric Resonator Antenna with Stair-Shaped Slot Excitation," *IEEE Trans. Antennas Propag.*, vol. 65, no. 3, pp. 1380–1383, 2016.
10. Y. Pan, K. W. Leung, and E. H. Lim, "Compact Wideband Circularly Polarised Rectangular Dielectric Resonator Antenna with Dual Underlaid Hybrid Couplers," *Microw. Opt. Technol. Lett.*, vol. 52, no. 12, pp. 2789–2791, 2010.
11. G. Varshney, V. S. Pandey, and R. S. Yaduvanshi, "Dual-Band Fan-Blade-Shaped Circularly Polarised Dielectric Resonator Antenna," *IET Microwaves, Antennas Propag.*, vol. 11, no. 13, pp. 1868–1871, 2017.
12. A. Rashidian and L. Shafai, "Compact Lightweight Polymeric-Metallic Resonator Antennas Using a New Radiating Mode," *IEEE Trans. Antennas Propag.*, vol. 64, no. 1, pp. 16–24, 2016.
13. A. Petosa and S. Thirakoune, "Rectangular Dielectric Resonator Antennas with Enhanced Gain," *IEEE Trans. Antennas Propag.*, vol. 59, no. 4, pp. 1385–1389, 2011.

4 Polarization in Dielectric Resonator Antenna

4.1 INTRODUCTION

The polarization of the radiated wave of the antenna in the far-field region is our main concern. The polarization of a radiated wave is the characteristics of an electromagnetic wave describing the time varying direction and relative magnitude of the electric field vector w.r.t. spatial coordinates. Polarization is the sense of electric field vector, as observed along the direction of propagation. Typically, this is measured in the direction of maximum radiation. There are three classifications of antenna polarization: linear, circular, and elliptical. Circular and linear polarizations are special cases of elliptical polarization. Typically, antennas will exhibit elliptical polarization to some extent. Polarization is indicated by the electric field vector of an antenna oriented in space as a function of time. Should the vector follow a line, the wave is linearly polarized. If it follows a circle, it is circularly polarized (either with a left-hand sense or right-hand sense). Any other orientation is said to represent an elliptically polarized wave. Circular polarization antenna can introduce diversity into reception. This can also provide frequency conservation. Circular polarization has the benefit of both horizontal and vertical polarization. Traditionally an antenna operates at a single frequency, where different types of antenna are used for different applications [1–3]. Circular polarization in antennas minimizes the sensitivity toward the misalignment between transmitting and receiving antennas. Hence, circular polarization in antennas allows us to mitigate the polarization mismatch losses due to the misalignment between the transmitting and receiving antennas. The problem of the signal multipath fading is removed by using the CP antennas, generation of polarization, using specific shape of the slot, using specific geometry of the DRA as radiator, and using dual/multi-feeding mechanisms in DRA. For example AM radio, FM radio and satellite radio have different operating frequencies for reception. Similarly, linear polarized and circular polarized signals have different patterns of propagation. Circular polarized signals can be termed robust signals as compared to linear signals. Similarly, using different antennas for different applications take more space and makes wireless communication systems complex. To solve this problem, multi-band response antenna can be used [4–7]. A dielectric resonator antenna, which is made up of a ceramic material of various shapes, is used for microwave frequency, and at higher frequencies, DRA is mounted on ground or metal surface. Radio waves are applied inside the DRA by using the transmitter circuit, and by bouncing of waves back and forth between the walls, they form standing waveform shape. The walls of the resonator are partially transparent to radio waves, which allows the radio waves to radiate in space. The dielectric resonator antenna is an omnidirectional antenna. An advantage of dielectric resonator antenna over other antennas is that they lack metal parts, which become lossy at high frequencies, giving lower conduction loss and providing more efficiency. The DRA size is inversely proportional to the under root of dielectric permittivity of material [8,9], so by increasing permittivity of material we can decrease the size of dielectric resonator antenna. Each mode of dielectric resonator antenna has a unique external and internal field (E-H) distribution. Therefore, different radiation patterns can be derived by exciting different modes of DRA. At present, DRAs are made up of plastic material (polyvinyl chloride (PVC)) [10–12]. DRA operates with fundamental mode in lower bands and with higher order mode in upper bands. This offers the upper 3-db axial ratio pass band with 10-db impedance pass band at different frequencies. The upper band response can be tuned with higher (2nd or 3rd) order mode by varying the position of the slot. Higher orders to fundamental mode ratios are 1.40 and 2.13 for different positions of slot. This ratio is acceptable for dual band response of the antenna [13,14]. A fan blade-shaped DRA is formed by using a rectangle and two half identical cylinder DRAs. A single micro-strip feed is provided to the DRA, which is less complex compared to dual feeding (Figure 4.1).

4.1.1 LINEAR POLARIZATION

In case of the linear polarization, the field vectors are arranged in a single direction, i.e., y-axis and plane wave propagate in the outward direction along the positive z-axis. The plane wave can be represented as

$E = E_1 \sin(\omega t - \beta z)$, where β is the propagation constant, E_1 is the amplitude of the wave. E_1 and E_2 are the amplitude of the linearly polarized signals.

Combining the above, the resultant field vector is given as:

$$E = \hat{x} E_1 \sin(\omega t - \beta z) + \hat{y} E_2 \sin(\omega t - \beta z + \delta) \tag{4.1}$$

$$E = \hat{x} E_1 \sin(\omega t) + \hat{y} E_2 \sin(\omega t + \delta)|_{z=0} \tag{4.2}$$

In case of the elliptical polarization, there are two components of the field in the x- and y-directions. The phase difference between them is δ. Both components of the field have different amplitudes $(E_1 \neq E_2)$. In this case, the resultant field vector rotates as a function of time. If the amplitude of both the field components is equal, then the wave is considered circularly polarized. The elliptically or circularly polarized waves can be represented as the combination of two linearly polarized waves (Figure 4.2).

$$E_x = E_1 \sin(\omega t - \beta z) \tag{4.3}$$

$$E_y = E_2 \sin(\omega t - \beta z + \delta) \tag{4.4}$$

At $z = 0$, $E_x = E_1 \sin \omega t$ and $E_y = E_2 \sin(\omega t + \delta) = E_2(\sin \omega t \cos \delta + \cos \omega t \sin \delta)$. Thus, solving these equations we get $\sin \omega t = \frac{E_x}{E_1}$ and $\cos \omega t = \sqrt{1 - (\frac{E_x}{E_1})^2}$.

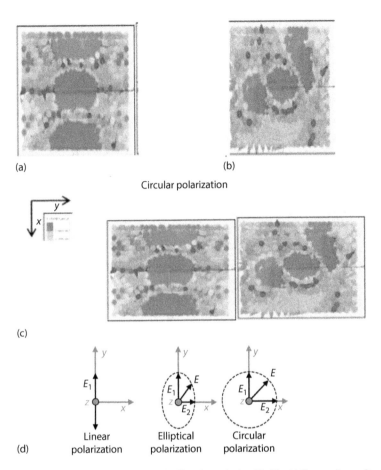

FIGURE 4.1 (a) Linear polarized fields. (b) Circular polarized fields. (c) Comparison of two fields (linear turned into circular polarized). (d) Representation of the different types of polarization.

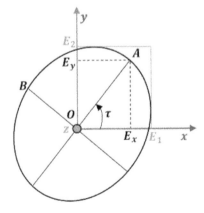

FIGURE 4.2 Polarization ellipse showing x and y components of the field with the peak values E_1 and E_2, respectively, and tilt angle τ.

Now solving these equations, we can get the equation representing the ellipse as given by

$$E_y = E_2 \frac{E_x}{E_1} \cos\delta + E_2 \sqrt{1 - \left(\frac{E_x}{E_1}\right)^2} \sin\delta$$

$$\left(\frac{E_y}{E_2} - \frac{E_x}{E_1} \cos\delta\right)^2 = \left(1 - \left(\frac{E_x}{E_1}\right)^2\right) \sin^2\delta$$

$$\frac{E_y^2}{E_2^2} + \frac{E_x^2}{E_1^2} \cos^2\delta - \frac{2E_x E_y \cos\delta}{E_1 E_2} = \left(1 - \left(\frac{E_x}{E_1}\right)^2\right) \sin^2\delta$$

$$\frac{E_y^2}{E_2^2} + \frac{E_x^2}{E_1^2} - \frac{2E_x E_y \cos\delta}{E_1 E_2} = \sin^2\delta \qquad (4.5)$$

In equation (4.3), all are constants except E_x and E_y. This equation represents the polarization ellipse. τ is the tilt angle of the ellipse.

1. For $E_1 = E_2$, $\delta = 0°$, linearly polarized wave with resultant arranged at angle 45°.
2. For $E_1 = 0$; LP wave along y-axis.
3. For $E_2 = 0$; LP wave along x-axis.
4. For $E_1 = E_2$; $\delta = +90°$; LHCP wave.
5. For $E_1 = E_2$; $\delta = -90°$; RHCP wave.

For a plane wave, the phase of the wave front remains constant, i.e., $\omega t - \beta z = \phi = $ constant. The plane wave is the function of z and t. This gives the relation for the function of time:

$$\omega = \frac{\beta dz}{dt} = \beta v_p \qquad (4.6)$$

where v_p is the phase velocity of the plane wave front in the $+z$ direction. The phase velocity of the plane wave is given as $v_p = 1/\sqrt{\mu\epsilon}$. For any medium $\mu = \mu_o$ and $\epsilon = \epsilon_r \epsilon_o$ give the relation $v_p = \frac{c}{\sqrt{\epsilon_r}} = \frac{c}{n}$, where ϵ_o is the permittivity, μ_o is the permeability of free space, ϵ is the permittivity of the medium, and ϵ_r and n are relative permittivity and refractive index of the medium, respectively.

If the phase is the function of position in space

$$\frac{d\omega}{dz} = \beta \qquad (4.7)$$

The Poynting vector of a circularly and elliptically polarized wave can be calculated as

$$S = \frac{1}{2}\left(E \times H^*\right) \tag{4.8}$$

The average Poynting vector is the real part of equation (1.36)

$$S_{av} = \frac{1}{2}Re\left(E \times H^*\right) = \frac{\hat{z}\left(E_1^2 + E_2^2\right)}{2Z_o} = \frac{\hat{z}E^2}{2Z_o} \tag{4.9}$$

$E_1^2 + E_2^2 = E^2$; E is the resultant field amplitude.

Linearly polarized antennas require accurate alignment to prevent polarization mismatch losses. Circularly polarized antennas are not limited by this factor. Spreading occurs in linearly polarized antennas due to multipath reception of signals. Circularly polarized antennas combat such problems.

The E-fields in DRA can be generated by proper excitation. When, $E_x = E_y$ in DRA, circular polarization is generated. Also if $E_x \neq E_y$; then elliptical polarization will be generated (Figures 4.3 through 4.6).

FIGURE 4.3 Anticlockwise rotation of fields, RHCP (right-hand circular polarization).

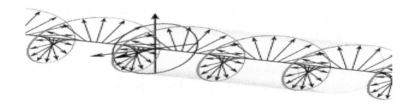

FIGURE 4.4 Clockwise rotating E-fields, LHCP (left-hand circular polarization).

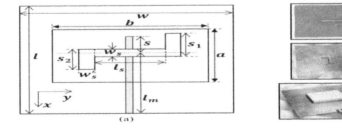

FIGURE 4.5 Circular polarization in DRA due to slot structure (a–d).

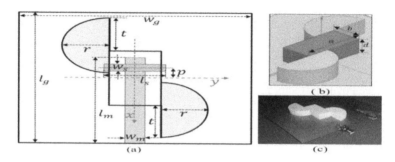

FIGURE 4.6   Circular polarization due to DRA (a–c).

4.1.2   Axial Ratio

The axial ratio (AR) is defined as the ratio of major axis to minor axis of an ellipse. From Figure 1.11, it is given by

$$AR = \frac{OA}{OB} \tag{4.10}$$

where:
   AR = 1; Circular polarization
   AR = ∞; Linear polarization

In general, this is difficult to achieve AR = 1. Hence, in the case $OB = \frac{OA}{\sqrt{2}} \rightarrow$ AR = 3 dB. We consider the criteria for obtaining the CP waves from an antenna. Misalignment of linear polarized antennas causes polarization mismatch losses. Multipath reception in linear polarized antennas leads to spreading, which causes degradation of the signal strength.. These problems can be minimized by circular polarized antennas. Signals transmitted by circular polarization can be received by both horizontal polarized as well as vertical polarized antennas. The polarization of waves is the arrangement of the field vectors in space. It can be categorized as linear, circular, and elliptical. The most general case of polarization is elliptical polarization. Axial ratio of linear polarizations is infinity and below 3 dB for circularly polarized DRAs. Axial ratio can be defined as ratio of major to the minor axis (Figures 4.7 and 4.8).

4.1.3   Circular Polarization

DRA dimensions: $r = 12$, $a = 24$, $b = 12$, $d = 10$, $t = 15$, $l_s = 13$, $= 1.5$, $l_m = 40$, $w_m = 1.6$, $s = 7$, $l_g = 80$, and $w_g = 80$ (All dimensions of antenna are in *mm*) (Figures 4.9 through 4.11). Figure 4.6a and b shows the proposed fan-blade shaped circularly polarized dielectric resonator antenna, which is placed above the ground plane with the ground plane above the substrate. The ground plane is rectangular with dimension ($l_s \times w_s$). The substrate is made up of FR_4 epoxy material with

FIGURE 4.7   DRA with circular polarization.

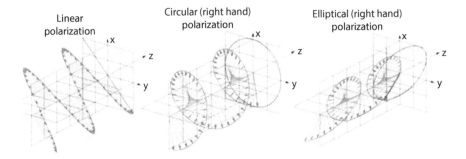

FIGURE 4.8   Three different wave fronts of linear, RHCP, and LHCP polarized signals.

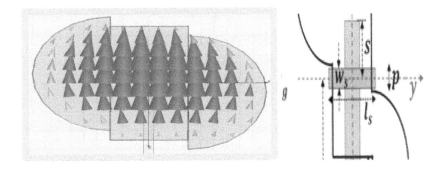

FIGURE 4.9   Linear polarization with rectangular slot as feed in DRA.

dimensions $l_g \times w_g$ with thickness 0.8 mm and permittivity $\epsilon_s = 4.4$. The radio frequency excitation is applied by using SMA connector and microstrip line with dimension ($l_m \times w_m$). Stub (s) is connected with microstrip line for impedance matching. DRA is simulated on high-frequency structure simulator (HFSS)-based on finite element method (FEM) (Figures 4.12 through 4.15).

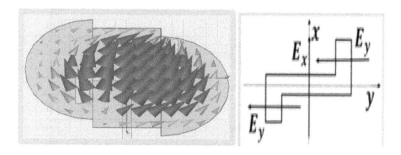

FIGURE 4.10   Circular polarization with Z-slot as feed in DRA.

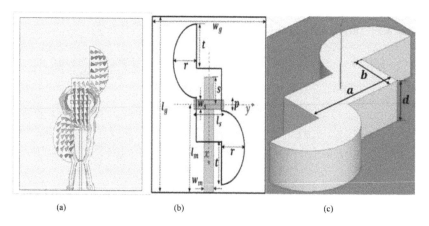

(a)   (b)   (c)

FIGURE 4.11   (a) E-field diagram of DRA. (b) Dielectric resonator antenna with feed. (c) 3D view of DRA.

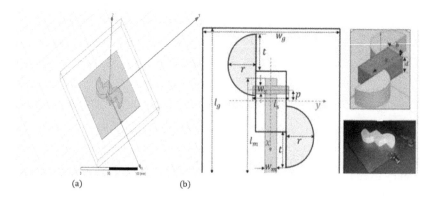

(a)   (b)

FIGURE 4.12   (a) Simulated DRA and (b) fabricated fan S-shaped circularly polarized dielectric resonator antenna.

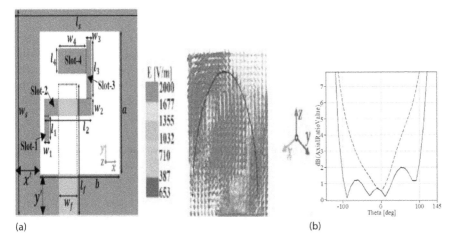

FIGURE 4.13 (a) DRA structure for circular polarization with E-field and (b) axial ratio.

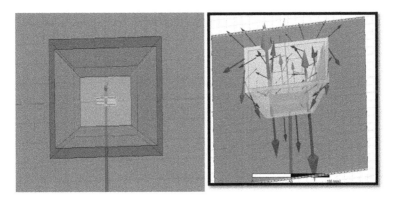

FIGURE 4.14 DRA horn rectangular E-fields without CP.

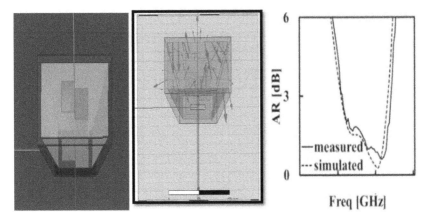

FIGURE 4.15 Horn DRA, E-fields with CP (with Z shape air gap in structure) with axial ratio plot.

4.2 RESULTS AND DISCUSSION

In this paper, the center of the antenna is assumed at the origin. The CPDRA is fabricated with slot position $p = 1.5$ mm on positive x-axis from the origin. Figure 4.4 shows the S_{11} parameter response versus frequency. Figure 4.4 shows the axial ratio and gain of the circularly polarized dielectric resonator antenna. The return loss $(S_{11}) < -10$dB of the proposed antenna has frequencies of 4.23, 7.23, and 8.21 GHz (Figures 4.16 and 4.17).

Figure 4.6a shows the E-field distribution of the DR at frequency 3.4 GHz when $t = 0$. The E-field distribution in DRA corresponds to the TE_{111} mode. Hence, DRA operates with the fundamental mode in lower band. Figure 4.6b shows the E-field distribution of the DRA at frequency 4.5 GHz; two half wave variation is observed along

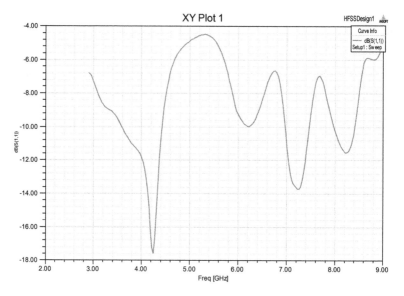

FIGURE 4.16   Simulated S_{11} of CPDRA.

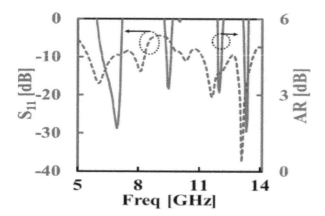

FIGURE 4.17   Axial ratio of CPDRA and S_{11}.

with the x-axis. Hence, the DRA operates with 2nd order mode (TE_{211}) in upper band. Figure 4.6c shows the H-field distribution in DRA at frequency 4.5 GHz, corresponding to the TM_{211} mode. The CPDRA operates with 2nd order mode in upper band. The E-field and H-field are at 90° phase difference. Figure 4.6a–c shows the E-field distribution of DRA at frequency 3.4 GHz when $t = T/4$ to show the circularly polarized response of antenna. A 90° phase difference can be observed between E-field vectors in the DRA. Placing half cylindrical DRAs at quarter wavelengths from the center of rectangular DRA as shown in the Figure 4.6b rotated e.m. fields because of circular polarization. Here, orthogonal mode pairs have been generated. The radiation pattern of the antenna can determine RHCP or LHCP depending on dominated direction of the field during rotations. The z shape feed also has introduced circular rotations into fields. The bigger arm length of z shape will introduce dominated fields. Hence, depending upon dominance of fields, RHCP/LHCP can be generated (Figures 4.18 through 4.21).

Radiation pattern of CPDRA with slot position $p = 1.5$ mm is studied and simulated on HFSS at different frequencies as shown in Figure 4.22; RHCP (right-hand circularly polarized) and LHCP (left-hand circularly polarized) are observed.

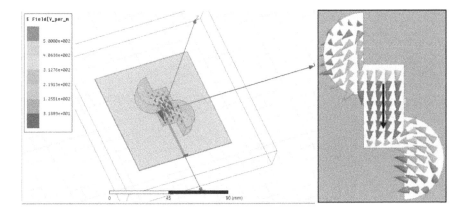

FIGURE 4.18 E-field distribution of DRA at frequency 3.4 GHz when $t = 0$.

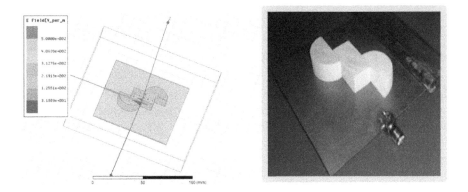

FIGURE 4.19 E-field distribution of DRA at frequency 4.5 GHz.

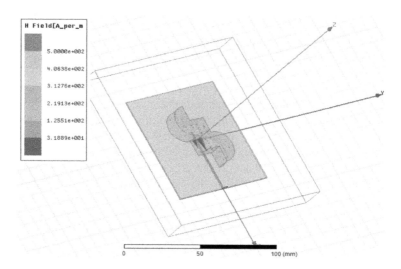

FIGURE 4.20   H-field distribution of DRA at frequency 4.5 GHz.

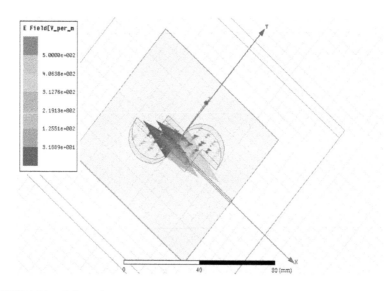

FIGURE 4.21   E-field distribution of DRA at frequency 3.4 GHz when $t = T/4$.

The field lines shows the LHCP and black solid line shows the RHCP. The DRA provides the right-hand circular polarization (RHCP) radiation in lower band and left-hand circular polarization (LHCP) in upper band. The polarization is reversed by changing the slot position from +1.5 to −1.5 mm. The polarization is reversed in upper band from LHCP to RHCP. Hence, slot is considered as the source of magnetic current. By changing slot position, the direction of magnetic current changes due to which polarization in reversed in upper band from LHCP to RHCP. The proposed antenna gives wider axial ratio and impedance bandwidths. The response of lower band (fundamental mode operation) of DRA remains the same in both positive and negative slot positions (Figure 4.22).

At fundamental mode of operating (in lower band) frequencies 3.4 and 3.72 GHz, RHCP dominates over LHCP. At 2nd order mode or higher order mode of operating frequencies (in upper band) 4.8 and 5.22 GHz, LHCP dominates over RHCP.

When slot position p is equal to or greater than 2.5 mm (in position x-axis direction), CPDRA operates in upper band with 2nd order mode at lower frequencies of C-band. When slot position p is equal and smaller than 0 mm (in negative x-axis direction) CPDRA operates in upper band with 3rd order mode at higher frequencies of C-band. Axial ratio and S_{11} response of the CPDRA remains the same in lower band for all (positive and negative) slot positions.

Dielectric resonator is made up of RT/duroid 6006/6010LM. The FR4_epoxy uses as the substrate with thickness of 0.8 mm and permittivity 4.4 (Table 4.1).

The frequency band over which AR remains less than or equal to 3 dB is called the AR bandwidth. Any antenna is considered circularly polarized if both imped-ance pass band $\left(S_{11} \le -10\,\mathrm{dB}\right)$ (Figures 4.23 through 4.25).

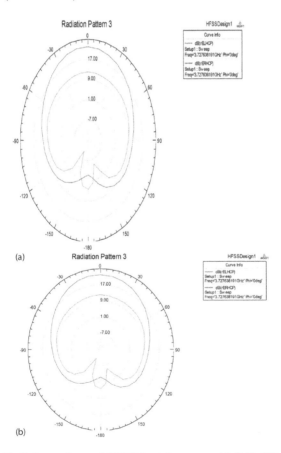

FIGURE 4.22 Radiation pattern of CPDRA at frequency (a) 3.42 GHz, (b) 3.72 GHz.
(*Continued*)

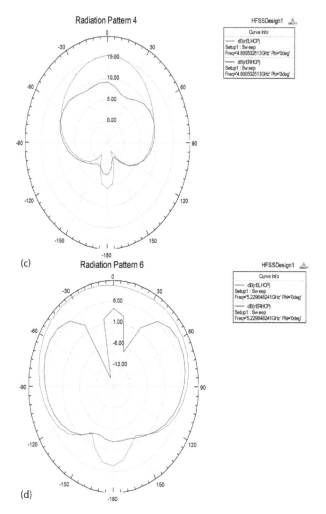

FIGURE 4.22 (Continued)   Radiation pattern of CPDRA at frequency (c) 4.8 GHz, and (d) 5.22 GHz.

TABLE 4.1
DRA Dimensions

Components	X-Size (mm)	Y-Size (mm)	Z-Size (mm)
Substrate	80	80	0.8
Ground	80	80	0
Rectangle (DRA)	24	12	10
Slot	1.5	13	0
Microstrip	47	1.6	0
Port	0	4	3
2-Identical Cylinders (DRA)	Radius = 12	Height = 10	

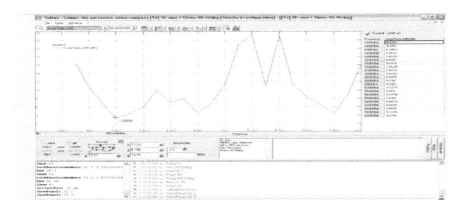

FIGURE 4.23 Measured axial ratio.

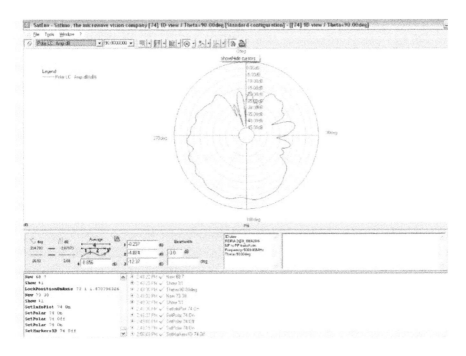

FIGURE 4.24 Measured radiation pattern LHCP.

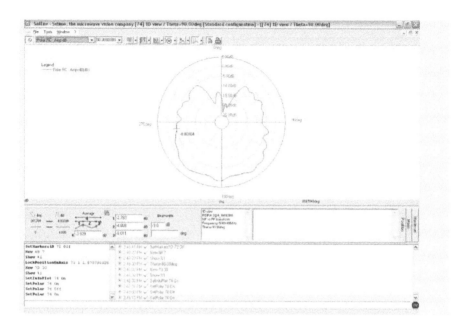

FIGURE 4.25   Measured radiation pattern RHCP.

4.3   CONCLUSION

Circular polarized in multiband antennas are in great demand for communication systems and vehicular networks communication. These can introduce diversity into systems. As per literature, multiband circular polarization has been achieved up to double bands only. Three or four bands with circular polarization have become challenging and futuristic work. DRA with fundamental and higher order modes excitations with special feed may result in multiband antennas with circular polarization. A single microstrip fed CPDRA, with dual band, circularly polarized response has been implemented. CP response of DRA can be tuned to either 2nd or 3rd order higher mode by changing the slot positions. The polarization of DRA upper band can be altered by changing its slot position. Reflection coefficients at frequencies of 4.23, 7.23, and 8.21 GHz have been obtained. Wider impedance bandwidths and axial ratio bandwidths have been achieved in lower and upper band in DRA. The dielectric resonator antenna is introduced with circular polarization. This is achieved by introducing electric RF current rotations. Circular polarization can also be achieved by slot and DRA structures. Implementation of circular polarization in single band antennas is easy. When it comes to multiband antennas, getting circular polarization becomes difficult and complex. Theories, experimentation, and simulations have been worked out to achieve different polarization into DRA.

REFERENCES

1. R. K. Mongia and P. Bhartia, "Dielectric Resonator Antennas—A Review and General Design Relations for Resonant Frequency and Bandwidth," *Int. J. Microw. Millimeter-Wave Comput. Eng.*, vol. 4, no. 3, pp. 230–247, 1994.
2. K. X. Wang and H. Wong, "A Circularly Polarized Antenna by Using Rotated-Stair Dielectric Resonator," *IEEE Antennas Wirel. Propag. Lett.*, vol. 14, pp. 787–790, 2015.
3. S. Fakhte, H. Oraizi, R. Karimian and R. Fakhte, "A New Wideband Circularly Polarized Stair-Shaped Dielectric Resonator Antenna," *IEEE Trans. Antennas Propag.*, vol. 63, no. 4, pp. 1828–1832, 2015.
4. G. Varshney, V. S. Pandey, R. S. Yaduvanshi and L. Kumar, "Wide Band Circularly Polarized Dielectric Resonator Antenna with StairShaped Slot Excitation," *IEEE Trans. Antennas Propag.*, vol. 65, no. 3, pp. 1380–1383, 2017.
5. X. S. Fang and K. W. Leung, "Linear-/Circular-Polarization Designs of Dual-/Wide-band Cylindrical Dielectric Resonator Antennas," *IEEE Trans. Antennas Propag.*, vol. 60, no. 6, pp. 2662–2671, 2012.
6. Y. Ding, K. W. Leung and K. M. Luk, "Compact Circularly Polarized Dualband Zonal-Slot/DRA Hybrid Antenna Without External Ground Plane," *IEEE Trans. Antennas Propag.*, vol. 59, no. 6 PART 2, pp. 2404–2409, 2011.
7. H. San Ngan, X. S. Fang and K. W. Leung, "Design of Dual-Band Circularly Polarized Dielectric Resonator Antenna Using a Higher-Order Mode," *Proc. IEEE-APS APWC*, pp. 424–427, 2012.
8. X. Fang, K. W. Leung and E. H. Lim, "Singly-Fed Dual-Band Circularly Polarized Dielectric Resonator Antenna," *IEEE Antennas Wirel. Propag. Lett.*, vol. 13, pp. 995–998, 2014.
9. Y. M. Pan, S. Y. Zheng and W. Li, "Dual-Band and Dual-Sense Omnidirectional Circularly Polarized Antenna," *IEEE Antennas Wirel. Propag. Lett.*, vol. 13, pp. 706–709, 2014.
10. M. Zhang, B. Li and X. Lv, "Cross-Slot-Coupled Wide Dual-Band Circularly Polarized Rectangular Dielectric Resonator Antenna," *IEEE Antennas Wirel. Propag. Lett.*, vol. 13, pp. 532–535, 2014.
11. Y. D. Zhou, Y. C. Jiao, Z. B. Weng and T. Ni, "A Novel Single-Fed Wide Dual-Band Circularly Polarized Dielectric Resonator Antenna," *IEEE Antennas Wirel. Propag. Lett.*, vol. 15, pp. 930–933, 2016.
12. X. Sun, Z. Zhang and Z. Feng, "Dual-Band Circularly Polarized Stacked Annular-Ring Patch Antenna for GPS Application," *IEEE Antennas Wirel. Propag. Lett.*, vol. 10, pp. 49–52, 2011.
13. L.-Y. Tsen and T.-Y. Han, "Circular Polarization Square-Slot Antenna for Dual-Band Operation," *Microw. Opt. Technol. Lett.*, vol. 50, no. 9, pp. 2307–2309, 2008.
14. R. S. Yaduvanshi. "Rectangular DRA design and theory," Springer, 2015.

5 Sapphire Dielectric Resonator Antenna

5.1 INTRODUCTION

Sapphire is a new candidate to be used as DRA that is embedded with robustness. It can be used in microwave as well as millimeter wave frequencies. The fabrication of sapphire antenna is simple. It has very good transmission properties. It is available in natural form and also can be developed. Its chemical compound formula is AL_2O_3. Its wafer as well as crystal forms are valuable in the open market.

There is a huge demand for highly efficient communication devices in terms of speed, size, bandwidth, beam width, and application-oriented designs. Military grade antennas have huge requirements. Sapphire dielectric resonator antennas can fulfill military demands. Investigation on the dielectric resonator antenna was first carried by Long et al. [1]. In the last few decades, DRA investigations have led to multiple studies and practical implementations. Dielectric resonator antenna is lossy at high frequencies and there are no conductor losses, therefore high radiation efficiency, large bandwidth, and flexible feed arrangement due to no metal parts [2,3]. That is why dielectric resonator antennas is preferred as compared to the conventional antenna. As per the literature review, DRA antenna, have been mainly realized by making use of ceramic materials characterized by high permittivity and high Q factor (between 2 and 2000). DRA is made from plastic material like polyvinyl chloride (PVC). The antenna designed from this material is not suitable for rugged and long distance communication applications. New material is needed to meet the next generation antenna requirements. Hence, sapphire can be used since it has a number of advantages such as durability and excellent transmission parameters. Sapphire can be used to design DRA antennas [4]. As per the literature survey, much less research work has been done on the sapphire-based antennas. In this chapter, various sapphire antennas are discussed such as stacked rectangular DRA, amalgamation of sapphire and TMM13i cuboids. The size of an antenna can also be reduced if DR material has high permittivity. The antenna size can also be easily varied by using sapphire structure with generation of higher order modes [14,15]. It is the best suitable candidate for future smart phones, laptops, iPods, and mobile communication.

This article presents a sapphire stacked rectangular dielectric resonator with aperture coupled feed. The designed antenna consists of a two-layer structure with TMM stacked over sapphire, as is shown in Figures 5.1 through 5.3.

The stacking structure has been used as this approach. The stacked dielectric resonator antenna proves to be an effective way to enhance gain and bandwidth as compared to single dielectric resonator antenna [5–7].

FIGURE 5.1   Sapphire antenna.

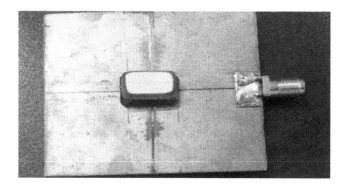

FIGURE 5.2   Sapphire antenna stacking with TMM.

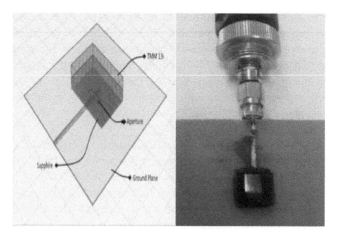

FIGURE 5.3   Sapphire antenna with feed.

The rectangular shaped DRA has fabrication flexibility in comparison to other geometries. This shape helps to achieve better radiation characteristics along with better impedance match [8–11].

There are various feed schemes for DRA like a microstrip transmission line, coaxial probe fed, coplanar waveguide line, and aperture coupled feed, to name a few. Aperture coupled feed has been used in this design, as it has the advantage of feeding networks kept below the ground plane and hence avoided the generation of spurious resonant modes [12,13]. This flow has been organized as follows: Introduction is covered in Section 5.1. Section 5.2 presents the mathematical modeling. The structure of design antenna is discussed in Section 5.3. Section 5.4 presents various obtained results and their discussion. A conclusion and future scope is presented in Section 5.5.

Next, possible designs using sapphire can be proposed with circular polarization. Sapphire antenna with multiple bands having features of circularly polarization, shall be very useful in mobile communications.

5.2 MATHEMATICAL MODELING

By using a dielectric waveguide model (DWM) resonant frequency, the initial dimensions of designed rectangular DRA are determined. By the magnetic wall boundary condition and solving the following transcendental equation, the resonator frequencies for dominant modes (TE to Z mode) are obtained; these are discussed below in equations 5.1 through 5.6. These equations are used to determine resonant frequency of antenna. Depending on boundary conditions, radiated fields are determined.

$$k_x^2 + k_y^2 + k_z^2 = \varepsilon_r k_0^2 \tag{5.1}$$

$$f_0 = \frac{c}{2\pi\sqrt{\varepsilon_r}}\sqrt{k_x^2 + k_y^2 + k_z^2} \tag{5.2}$$

$$k_x = \frac{\pi}{a}; k_y = \frac{\pi}{b}$$

$$k_z\left(\tan\frac{k_z d}{2}\right) = \sqrt{(\varepsilon_r - 1)k_0^2 - k_z^2} \tag{5.3}$$

where ε_r denotes the dielectric constant of dielectric resonator antenna, c is the velocity of light, and k_0, k_x, k_y and k_z are wave numbers along x, y, and z directions, respectively.

Since the aperture coupled feed mechanism has been used, the following equations are used as shown in Figure 5.4.

Slot length,

$$L_S = \frac{0.4\lambda_o}{\sqrt{\varepsilon_e}} \tag{5.4}$$

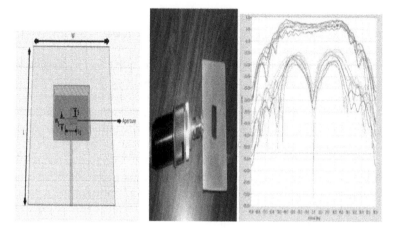

FIGURE 5.4   Aperture coupled sapphire DRA.

where λ_o is the wavelength and the effective permittivity is defined as

$$\varepsilon_e = \frac{\varepsilon_r + \varepsilon_s}{2}$$

where ε_r and ε_s are the dielectric constant of the rectangular dielectric resonator and substrate, respectively.

Slot width,

$$W_s = 0.2L_s \qquad\qquad (5.5)$$

And the stub length,

$$s = \frac{\lambda_g}{4} \qquad\qquad (5.6)$$

where λ_g is the guided wave in the substrate.

5.3   SAPPHIRE ANTENNA DESIGN

The sapphire antenna design consists of two stacked rectangular dielectric resonator antenna fabricated on FR4 substrate of dielectric constant 4.4 and dimensions 50×50 mm^2 with the loss tangent 0.002. The sizes of both rectangular dielectric resonators used for the design are made up of different materials and dimensions as well as permittivity. The Saphire antenna is stacked with two different materials layers. Top layer is saphire and Lower layer is TMM. The lower dielectric resonator antenna of height 2.5 mm is made of sapphire having a dielectric constant of 10, and the upper dielectric resonator antenna of height 2.5 mm is made up of TMM13i, which is thermoset microwave material with dielectric constant of 12.8

as shown in Figure 5.3. Sapphire has the advantage of corrosion resistance and better radiation resistance. It has the lowest losses by light absorption and better scattering characteristics. Sapphire can be used in medical applications such as surgical systems for laser transmission and optical applications. Sapphire can be used for short and long wavelength applications at IR [16]. Cryzal and Ellisor have developed concave sapphire lenses that can be used as superstrate for beam width control of antenna [11–16]. Their popular applications are to increase focal lengths in existing systems. M/s Luoyang Dingming Optical Technology Co., Ltd. is a Chinese material supplier, providing optoelectronics and semiconductor substrate to design sapphire antennas. M/s DM Luoyang Dingming Optical Technology Co., Ltd. can develop sapphire products. Concave lenses decreased the beam width in antenna if placed as sapphire lens (Cryzal or Ellisor). Convex lenses work opposite for beam width. DM Luoyang Dingming Optical Technology Co., Ltd. product development. Sapphire material is good candidate for antenna applications. Zircar (Al_2O_3) is another dielectric material used for DRA designs. Single sapphire crystals are used as DRA and NDRA (sizes available are 2 inch to 8 inch dimensions from M/s KYOCERA (Fine Ceramics)). ZIRCAR is also available in the open market. The application of sapphire antenna can be used as DRA beam control as well as beam width control (Figures 5.5 through 5.8).

For the purpose of providing the feed arrangement, a feed line of 1.2 mm × 30.5 mm and a slot of 1.5 mm × 9 mm has been etched from the ground plane to create the aperture coupling excitation mechanism shown in Figure 5.8.

All the design dimensions are summarized in Table 5.1 and Figure 5.9.

FIGURE 5.5   Green emerald (panna) antenna for futuristic antenna usage.

FIGURE 5.6   Coral as possible use of DRA. Coral Moga (calcium carbonate) dielectric constant is 6.1 to 9.1.

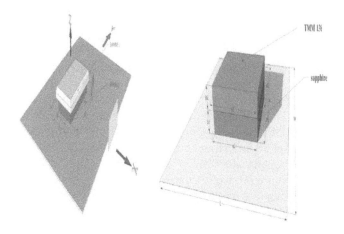

FIGURE 5.7   Sapphire antenna.

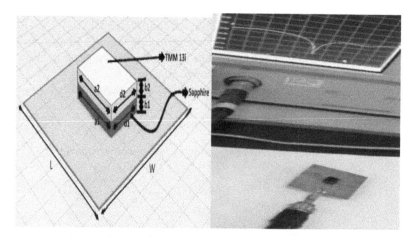

FIGURE 5.8   Geometry of SDRA. $L = 50$ mm, $W = 50$ mm, $a_1 = 13$ mm, $d_1 = 10$ mm, $b_1 = 2.5$ mm, $a_2 = 12$ mm, $d_2 = 8$ mm, and $b_2 = 2.5$ mm.

TABLE 5.1
Design Parameters of Sapphire DRA

Object	Material	Dielectric Constant (ε_r)	Dimension Specifications in mm		
			Length	Width	Height
Ground Plane	FR4	4.4	50(L)	50(W)	0.8
Dielectric Layer 1	Sapphire	10	13(a_1)	10(d_1)	2.5(b_1)
Dielectric Layer 2	TMM13i	12.8	12(a_2)	8(d_2)	2.5(b_2)

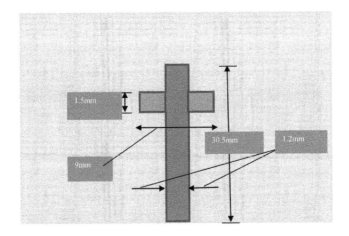

FIGURE 5.9   Feed (top view).

5.4   SAPPHIRE ANTENNA RESULTS AND DISCUSSION

5.4.1   Simulated Result

The designed antenna has been simulated using HFSS. Measured results of sapphire antenna have been obtained at 7 GHz. The reflection coefficient (S_{11}) parameter response of the antenna is shown in Figure 5.10. The frequency band where $S_{11} < -10$ dB is known as impendence bandwidth. Sapphire antenna can also be designed to obtain circular polarization using a z-shape feed or z-shape DRA. It is observed that antenna exhibits <10 dB reflection over the frequency band from 6.81 to 7.24 GHz.

The impedance matching is one of the most important criteria for antenna design. Figure 5.11 shows 50 ohms impedance matching. The criteria of $S_{11} < -10$ dB and VSWR less than or equal to 2 show a good impedance matching of an antenna. The VSWR was found to be under 2 as shown in Figure 5.12. The radiation patterns at Phi-0deg and Phi-90deg are shown in Figures 5.13 and 5.14, respectively.

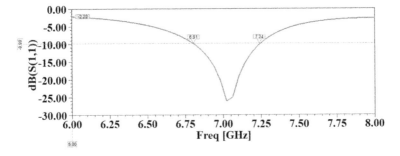

FIGURE 5.10 $|S_{11}|$ return losses vs frequency plot.

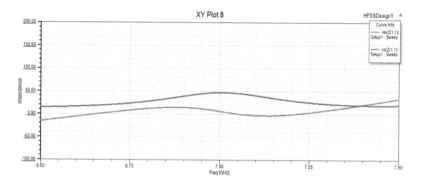

FIGURE 5.11 $|Z_{11}|$ impedance vs frequency plot.

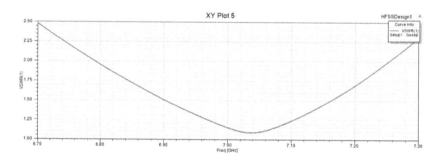

FIGURE 5.12   VSWR vs frequency plot.

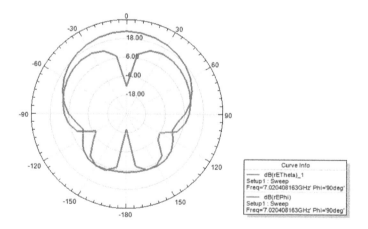

FIGURE 5.13   Radiation pattern at Phi-0deg.

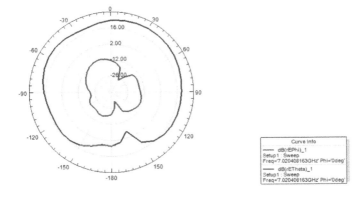

FIGURE 5.14   Radiation pattern at Phi-90deg.

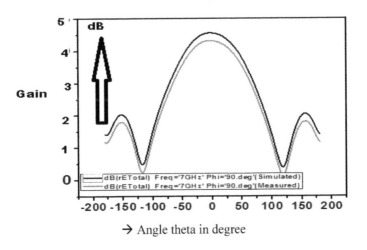

→ Angle theta in degree

FIGURE 5.15   2D Radiation pattern at 7 GHz, along *x*-axis and gain in dB along *y*-axis.

The obtained gain of 5.2 dB is shown in Figure 5.15. Hence the antenna has been perfectly matched before fabrication. Results of perfect match can be obtained by an impedance match plot or smith chart. Radiation has been observed with 2D radiation plots. These can be verified by simulators as well as anechoic chambers and VNA measured results. If both results are found, same then design shall be appreciated. Figure 5.15 shows that beam width control can be obtained using higher order modes or by possible use of superstrate in DR antennas.

5.4.2 Fabrication and Measured Results of Sapphire DRA

The prototype of the antenna was fabricated as shown in Figure 5.16. The various tested results on VNA (vector network analyzer) and anechoic chamber are presented and discussed.

The various experimental results of the designed antenna such as return loss, gain, and radiation pattern are shown in Figures 5.11 through 5.15. The simulated and measured 10 dB impedance bandwidth of the antenna is 6.81 to 7.24 GHz and 6.79–7.27 GHz, shown in Figure 5.17. The radiation patterns of the antenna in

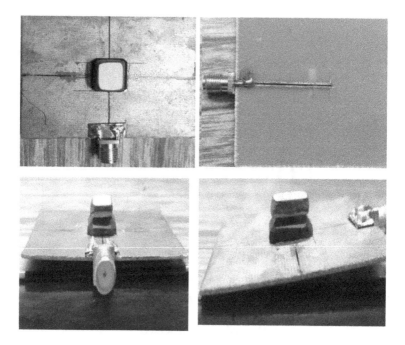

FIGURE 5.16 Photographs of the fabricated stacked antenna top view/back view, side views.

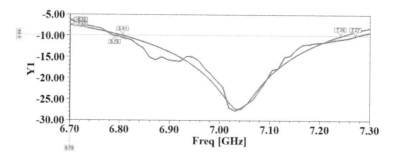

FIGURE 5.17 Measured result and simulated result of $|S_{11}|$ return losses vs frequency plot.

TABLE 5.2

Comparison of the Proposed SDRA with Existing DRAs

References	Material	Dielectric Constant (ε_r)	Thermal Conductivity (W/m·k)	Dissipation Factor (δ)	Frequency Achieved (GHz)
[14]	Roger TMM10i	9.8	0.76	0.0020	5.5–5.9
[15]	Soda Lime Glass	7.75	0.937	0.0005	5–6.3
[16]	Arlon 25N	3.38	0.45	0.0025	5.6–5.95
[17]	Rogers RT5880LZ	2	0.33	0.0021	5.21–6.84
Proposed Design	**Sapphire**	**10**	**27**	**0.0001**	**6.79–7.27**

the *xz* and the *yz* plane are shown in Figure 5.13. Figure 5.15 is a gain plot. In the observed frequency band, the radiation pattern of the antenna is broadside and stable. The experimental results are in close proximity of simulated results.

As per the literature survey, a comparative analysis of proposed DRA with other DRAs is summarized in Table 5.2.

5.5   CONCLUSION

An aperture coupled novel aesthetic sapphire DRA (sapphire dielectric resonator antenna) has been designed, fabricated, and experimented. The antenna has achieved a gain of 5.2 dB, voltage standing wave ratio (VSWR < 2). The simulated and measured −10 dB impedance bandwidth of the antenna is frequency range 6.81 to 7.24 GHz and 6.79–7.27 GHz, respectively, with impedance matching to 50 Ohms. The physical attributes of sapphire-like hardness and stability to chemical erosion make it a suitable candidate for long life, underground use, water use, and rugged communication. These sapphire antennas can also be used for 5G applications. The antenna size can further be miniaturized for mobile communication. Hence, it can be used for future smart phones. The future extension of work can be taken up for designing the sapphire antennas in THz ranges embedded with circular polarization features. Sapphire is an alumina (Al_2O_3) material. The higher order modes can be generated in sapphire antennas. Wide band design can also be obtained using merged modes. These higher order modes can be generated by shifting feed position and proper selection of aspect ratio in between 0.5 and 2.5. Using sapphire concave lens as superstrate, antenna beam width control can be achieved. This narrow beam width generation and control has excellent applications for vehicular antennas and smart cities. Sapphire with rectangular/ concave/convex type superstrate in DRA can be used for beam width control.

REFERENCES

1. [a] Elena R. Dobrovinskaya, Leonid A. Litvinov and Valerian Pishchik, *Sapphire Material, Manufacturing, Applications*. New York: Springer Publishing, 2009; [b] Nanjing Sapphire Electro-Optics Co., Ltd. No. 22 N. Liuzhou Rd., Nanjing Hi-Tech Development Zone, Nanjing 210031, P.R. China, market@mpasapphire.com; [c] S.A. Long, M. W. McAllister and L. C. Shen, "The Resonant Cylindrical Dielectric Cavity Antenna," *IEEE Trans. Antennas Propag.*, vol. 31, pp. 406–412, 1983.

2. B. Li and K. W. Leung, "A Wideband Strip-Fed Rectangular Dielectric Resonator Antenna," *IEEE Trans. Antennas Propag.*, vol. 53, no. 7, pp. 2200–2207, 2005.

3. Y. Zhang, J.-Y. Deng, M.-J. Li, D. Sun and L.-X. Guo, "A MIMO Dielectric Resonator Antenna with Improved Isolation for 5G mm-Wave Applications," *IEEE Antennas Wirel. Propag. Lett.*, pp. 1–1, 2019.

4. G. Bakshi, R. Singh Yaduvanshi and A. Kumar, "Design of Rectangular Dielectric Antenna Using Sapphire with Probe Fed Arrangement," *IJIEEE*, vol. 4, no. 9, 2016.

5. Y. M. Pan and S. Y. Zheng, "A Low-Profile Stacked Dielectric Resonator Antenna with High-Gain and Wide Bandwidth." *IEEE Antennas Wirel. Propag. Lett.*, vol. 15, pp. 68–71, 2015.

6. A. A. Kishk, X. Zhang, A. W. Glisson and D. Kajfez, "Numerical Analysis of Stacked Dielectric Resonator Antennas Excited by a Coaxial Probe for Wideband Applications," *IEEE Trans. Antennas Propag.*, vol. 51, no. 8, pp. 1996–2006, 2003.

7. A. Petosa, N. Simons, R. Siushansian, A. Ittipiboon and M. Cuhaci, "Design and Analysis of Multi-segment Dielectric Resonator Antennas," *IEEE Trans. Antennas Propag.*, vol. 48, no. 5, pp. 738–742, 2000.

8. B. Li and K. W. Leung, "Strip-Fed Rectangular Dielectric Resonator Antennas with/ Without a Parasitic Patch," *IEEE Trans. Antennas Propag.*, vol. 53, no. 7, 2005.

9. R. K. Mongia and A. Ittipiboon, "Theoretical and Experimental Investigations On Rectangular Dielectric Resonator Antennas," *IEEE Trans. Antennas Propag.*, vol. 45, pp. 1348–1356, 1997.

10. M. H. Neshati and Z. Wu, "Rectangular Dielectric Resonator Antennas: Theoretical Modelling and Experiments," *7th International Conference on Antennas and Propagation*, April 17–20, 2001, Conference Publication No. 480 IEEE 2001.

11. S. Maity and B. Gupta, "Theory and Experiments on Horizontally Inhomogeneous Rectangular Dielectric Resonator Antenna," *AEU-Int. J. Electron. Commun.*, vol. 76, pp. 158–165, 2017.

12. J. St. Martin, Y. M. M. Antar, A. A. Kishk, A. Ittipiboon and M. Cuhaci, "Dielectric Resonator Antenna Using Aperture Coupling," *Electron. Lett.*, vol. 26, no. 24, pp. 2015–2016, 1990.

13. J. St. Martin, Y. M. M. Antar, A. A. Kishk, A. Ittipiboon and M. Cuhaci, "Aperture-Coupled Dielectric Resonator Antenna", *IEEE International Symposium on Antennas and Propagation Digest*, pp. 1086–1089, 1991.

14. A. Z. Ashoor and O. M. Ramahi, "Dielectric Resonator Antenna Arrays for Microwave Energy Harvesting and Far-Field Wireless Power Transfer," *Prog. Electromagn. Res. C*, vol. 59, pp. 89–99, 2015.

15. A. A. Masius, Y. C. Wong and K. T. Lau, "Miniature High Gain Slot-Fed Rectangular Dielectric Resonator Antenna for IoT RF Energy Harvesting," *Int. J. Electron. Commun.*, vol. 85, pp. 39–46, 2017.

16. M/s. KYOCERA India Pacific Asia Pvt Ltd., India, Haryana Guru gram Road. G.M, Square Regent JMD, Floor 10th, 1004B & 1004A India, phone – 0124-4714298.

17. A. A. Masius, and Y. Wong, "Design of high gain co-planar waveguide fed staircase shaped monopole antenna with modified ground plane for RF energy harvesting application," *Proceedings of Mechanical Engineering Research Day*, pp. 118–119, 2017.

6 Miniaturization and De-miniaturization of DRA

6.1 INTRODUCTION

Nanotechnology is in demand because of compact devices. Device dimensions are reduced from centimeters to nanometer lengths. Biomedical sensors and robots due to nanotechnology have become possible for taking images of inner parts of the human body. Nano antenna can be embedded into humans for monitoring heartbeat, blood pressure, and other parameters remotely. Hence, the size of communication devices is being reduced as technology development takes place day by day. For wireless communication, the antenna is the key element, and 5G communication requires high data rates and high gain. High bandwidth antenna can support communication systems for increased bandwidth [1]. Shifting to sub millimeter wave frequencies has become today's requirement [2]. DRA has become an ideal candidate to get wider [3]. Higher modes can be generated in DRA [4], and merging of higher order modes can provide large bandwidth in DRAs [5].

Polarization of an antenna has different roles to play [6]. Circular polarization can avoid the problem of misalignments of antennas [7]. Initially, only linearly polarized (LP) DRAs were developed [8]. The LP antennas have the problem of misalignment and polarization mismatch losses, and these are being replaced by circular polarized antennas [9]. The Circular polarization is achieved by feeding mechanism or structure or both [10]. Use of high permittivity can also reduce the size of DRA [11]. High gain of antenna can also be obtained by introducing the concept of operating DRA at higher order modes. To overcome this problem, miniaturization techniques have been proposed to reduce the size of the DRA for a constant resonant frequency [12]. Generally, increase in height of the DRA can increase gain of the DRA due to generation of higher order modes [13].

The resonant frequency of the antenna at fundamental is shifted to the lower side.

6.2 ANTENNA CONFIGURATION

Figure 6.1 depicts a DRA with aperture coupled structure with a substrate of FR_4 epoxy of relative permittivity.

The thickness of the substrate is 0.8 mm. Slot is made in the ground plane. The fields are coupled to the DRA from microstrip line. A stub of desired length " " is connected to the microstrip line for impedance matching. The DRA is made up of the dielectric slabs of material with relative permittivity. The height of the DR is kept for fundamental mode at frequency around 5.3 GHz.

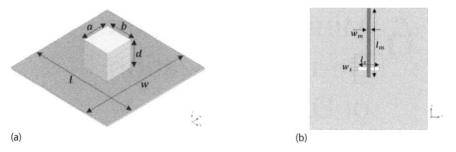

(a)   (b)

FIGURE 6.1   (a) DRA and (b) Feed.

6.3   RESULTS AND DISCUSSIONS

Figure 6.2a S_{11} are reflection coefficients with very simple aperture of slot coupled DRA made to operate at fundamental mode at frequency 5.23 GHz.

The second structure is coated with metallic nature. The height of the DRA is selected as per desired aspect ratio. After coating the top layer of DRA with metal characteristics, the resonant frequency is lowered to 3.90 GHz as shown in Figure 6.2a.

The E-field distribution inside the DRA loaded with metallic patch is shown in Figure 6.3. In the third antenna structure, the metallic patch is introduced at height of the DRA. This height of the metallic patch creates a further shift of the resonant frequency of the antenna. Introduction of metallic patch effects resonant frequency of the antenna, it changes to 3.70 GHz. Now, the resonant frequency of the antenna becomes 3.50 GHz as shown in Figure 6.2a.

The field distribution inside first antenna without metallic coating shown in Figure 6.3a at frequency 5.23 GHz is identical to the mode The field distribution of the fundamental mode is changed interestingly due to the presence of the metallic coating at different heights of the DR. Figure 6.3b–d shows the change in the field. The main observations of the Figure 6.3a–d reveal that in the portion of the DRA below the metallic patch, the E-field remains vertically distributed while the inverted

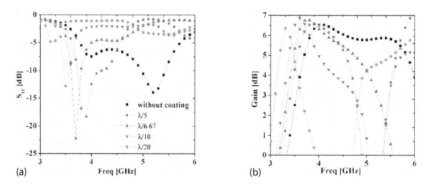

FIGURE 6.2   Simulation results (a) S_{11} and (b) gain of DRA.

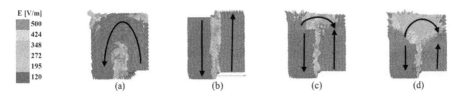

FIGURE 6.3 E-fields of DRA (a) 5.23 GHz, without metallic coating, (b) 3.90 GHz with patch, (c) 3.70 GHz with patch, and (d) 3.50 GHz with patch.

dipole is formed inside the portion of the DR that is above the metallic patch. However, the strong field remains below the metallic patch. Furthermore, the peak gain of the antenna remains around 7 dBi in all the cases. If the dielectric slabs above the metallic patch are removed in the third, fourth and fifth antennas, the gain of the antenna is significantly reduced. Figure 6.4 shows the radiation patterns of the different antenna structures without and with the metallic patch at different heights of the DRA. It can be observed that the radiation pattern of the antenna remains broadside as it is operating in fundamental mode. The level of cross polarization is quite low in bore sight direction; the cross polarization separation remains 50 dB or more (Figure 6.5).

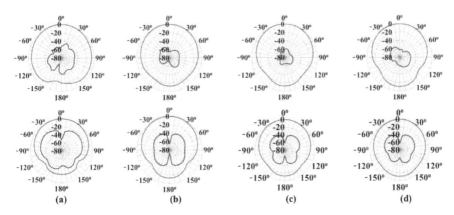

FIGURE 6.4 Radiation pattern (a) 5.23 GHz, without metallic coating, (b) 3.90 GHz with patch radiation pattern at, (c) 3.70 GHz with patch at, and (d) 3.50 GHz with patch at (top images in and bottom images in plane).

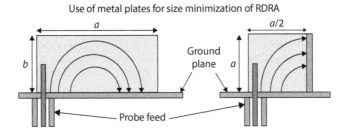

FIGURE 6.5 Image effect implementation.

6.4 CONCLUSION

Use of defected ground structure (DGS) and metal strips at a particular position is another technique to be developed. A miniaturization technique of rectangular DRA has been developed. A metallic patch is used to miniaturize the rectangular DRA. The resonant frequency of the rectangular DRA is shifted in the backward direction by changing the height of the metallic patch inside the DRA. The minimum height of the metallic patch is $\lambda/10$. Interestingly, the field configuration of the fundamental mode is changed by applying the miniaturization technique presented. Nanotechnology has evolved as an excellent technique to reduce the size of DRAs. Developing a hardware requires precision fabrication techniques. This an area of challenge to deal with miniaturization. Use of SRR (surface ring resonator), dumbbell DGS, routing of metal strips inside the structure and introduction of image effects with met material applications have been helpful to reduce the size of DRA.

REFERENCES

1. A. Goldsmith, *Wireless Communications*. Cambridge, MA: Cambridge University Press, 2005.
2. D. M. Pozar, *Microwave Engineering*. 4th ed. Hoboken, NJ: John Wiley & Sons, 2005.
3. R. S. Yaduvanshi and H. Parthasarathy, *Rectangular Dielectric Resonator Antenna Theory and Design*. New York: Springer US, 2016.
4. A. Petosa and A. Ittipiboon, "Dielectric Resonator Antennas: A Historical Review and the Current State of the Art," *IEEE Antennas Propag. Mag.*, vol. 52, no. 5, pp. 91–116, 2010.
5. G. Varshney, R. S. Yaduvanshi, and V. S. Pandey, "Gain and Bandwidth Controlling of Dielectric Slab Rectangular Dielectric Resonator Antenna," *12th IEEE Int. Conf. Electron. Energy, Environ. Commun. Comput. Control (E3-C3), INDICON 2015*, vol. 3, pp. 6–9, 2016.
6. T. Chang and J. Kiang, "Bandwidth Broadening of Dielectric Resonator Antenna by Merging Adjacent Bands," *IEEE Trans. Antennas Propag.*, vol. 57, no. 10, pp. 3316–3320, 2009.
7. G. Varshney, P. Mittal, V. S. Pandey, R. S. Yaduvanshi, and S. Pundir, "Enhanced Bandwidth High Gain Micro-Strip Patch Feed Dielectric Resonator Antenna," *Int. Conf. Comput. Commun. Autom.*, vol. 1, pp. 92–95, 2016.
8. R. K. Mongia and P. Bhartia, "Dielectric Resonator Antennas—A Review and General Design Relations for Resonant Frequency and Bandwidth," *Int. J. Microw. Millimeter-Wave Comput. Eng.*, vol. 4, no. 3, pp. 230–247, 1994.
9. G. Varshney, V. S. Pandey, R. S. Yaduvanshi, and L. Kumar, "Wide Band Circularly Polarized Dielectric Resonator Antenna with Stair-Shaped Slot Excitation," *IEEE Trans. Antennas Propag.*, vol. 65, no. 3, pp. 1380–1383, 2016.
10. Y. Pan, K. W. Leung, and E. H. Lim, "Compact Wideband Circularly Polarised Rectangular Dielectric Resonator Antenna with Dual Underlaid Hybrid Couplers," *Microw. Opt. Technol. Lett.*, vol. 52, no. 12, pp. 2789–2791, 2010.

11. G. Varshney, V. S. Pandey, and R. S. Yaduvanshi, "Dual-Band Fan-Blade-Shaped Circularly Polarised Dielectric Resonator Antenna," *IET Microwaves, Antennas Propag.*, vol. 11, no. 13, pp. 1868–1871, 2017.
12. A. Rashidian and L. Shafai, "Compact Lightweight Polymeric-Metallic Resonator Antennas Using a New Radiating Mode," *IEEE Trans. Antennas Propag.*, vol. 64, no. 1, pp. 16–24, 2016.
13. A. Petosa and S. Thirakoune, "Rectangular Dielectric Resonator Antennas with Enhanced Gain," *IEEE Trans. Antennas Propag.*, vol. 59, no. 4, pp. 1385–1389, 2011.

7 Hybrid Modes Excitation into DRA

7.1 INTRODUCTION

DRA can be excited using a probe or microstrip line with frequency based on dimensions and field propagating according to boundary conditions. The solution of electric and magnetic currents can be found using Maxwell's equation at resonant frequency. The electric charge densities and magnetic charge densities inside DRA (dielectric resonator antenna) are equated as per the law of conservation on top and bottom surfaces. On side walls normal components get vanished. H_z fields are obtained by linear combinations of sinusoidal functions. The direction of fields is to be taken in the x-y direction w. r. t. to source as z-components. Solving the Helmholtz equation for H_z, we get a second order linear differential equation. We obtain coefficient functions H_z.

Helmholtz equations for the EM fields with vector sources is determined. These are solved taking account from gradient and curl of the electric charge, magnetic charge, and current densities. We need to understand that tangential components of the magnetic field vanish on side walls.

Following is the complete solution for resonant modes excitation. H_z at the top and bottom surfaces is taken as, i.e., at $z = 0$, d. The normal components of the E-field vanish on the side walls. The e-m field inside DRA is proportional to $\frac{1}{\delta}$, where δ is the frequency perturbations determined from Dirac delta functions. The complete solution of resonant modes in DRA is thus achieved. Hybrid modes are superposition of TE and TM modes inside DRA [1–26].

In this hybrid, modes have been generated by using a probe of length d inserted into DRA in the z direction. The copper plane (x, y) of DRA is used to have current density. This is solved based on the KAM (Kalmogorov Arnold Moser) time averaging method and using δ-Dirac delta functions. The principle of orthogonality is finally applied to determine C_{mnp} and d_{mnp} amplitude coefficients. Here, H_z and E_z field have been computed simultaneously. The hybrid modes fields have been shown in Figures 7.1 and 7.2.

High efficiency and polarization diversity are achieved by hybrid modes along with conservation of frequency. The Helmholtz equation provided a solution of eigen functions. The longitudinal components E_z and H_z form transverse components of E_x, E_y, H_x, H_y.

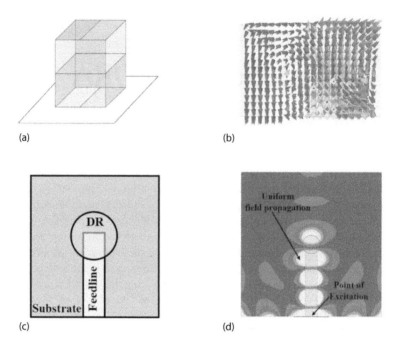

(a)   (b)

(c)   (d)

FIGURE 7.1 (a) Rectangular DRA with hybrid boundaries of PMC and PEC, (b) hybrid mode HEM$_{11\delta}$, (c) DRA hybrid mode excitation, and (d) hybrid mode fields.

A probe of length d is inserted from the ground plane along the z-direction. Surface current density is produced in DRA due input excitation. The feed point is located at $\left(\frac{a}{2},\frac{b}{2},z\right)$ of DRA. The azimuthally, ϕ-component of magnetic fields inside the resonator is introduced, and it corresponds to z-part of surface current density. The longitudinal components are E_z and H_z.

The current distribution inside DRA controlled resonant modes inside DRA. The inner product or reaction term of eigen function shall be equal to the corresponding eigen mode. The magnetic current is equal to electric currents in DRA. It maintains orthonormality or conservation of energy always.

Both, input excitation frequency and DRA radiated frequency are same. Extracting a particular resonant by large surface current density is possible due to input excitation. The desired coefficient becomes large for a particular mode, and it acts as the dominant mode. Higher gain and high directivity in DRA can thus avoid electromagnetic pollution (Figure 7.3).

The transverse components E_x, E_y, H_x, H_y are the components determined in terms of longitudinal components E_z, H_z. These transverse fields satisfy Helmholtz equations.

These are expressed as $u(mnp)\, e^{j\omega(mnp)t}$ and $v(mnp)\, e^{j\omega(mnp)t}$. E_x, E_y, H_x, H_y fields are partial derivatives of $u(mnp)$, $v(mnp)$. The $C(mnp)$, $D(mnp)$ are linear combinational coefficients of $u(mnp)$, $v(mnp)$ for E_z, H_z. These amplitude coefficients

$C(mnp)$, $D(mnp)$ can be determined by matching H_x, H_y at $z = 0$ to the surface current density of DRA when feed at $z = 0$.

If the surface excitation at $z = 0$ has a frequency component other than $\omega(mnp)$, say ω, then the fields amplitude components corresponding to this excitation are determined by the KAM theory of averaging; the DRA extracts out only $\omega(mnp)$ frequencies with amplitude.

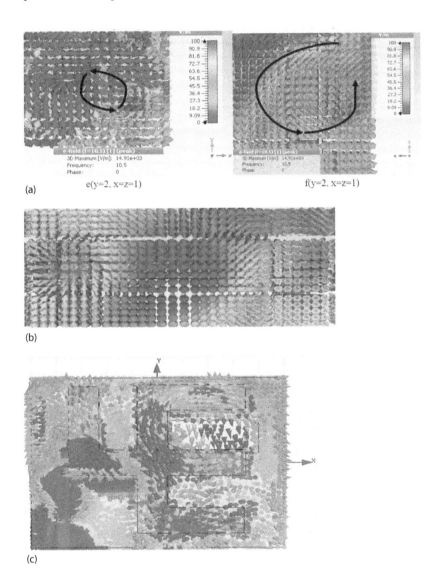

(a) $e(y=2, x=z=1)$ $f(y=2, x=z=1)$

(b)

(c)

FIGURE 7.2 (a) Hybrid mode TEM fields, (b) hybrid mode TEM fields, (c) hybrid mode TEM fields.   (*Continued*)

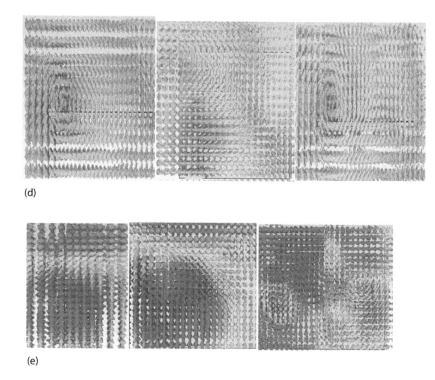

(d)

(e)

FIGURE 7.2 (Continued)   (d) *H* field vector, and (e) *E* field vector.

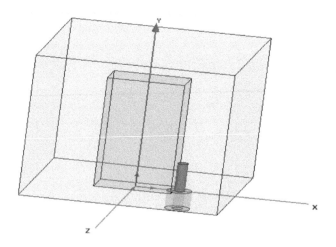

FIGURE 7.3   Rectangular DRA excited with probe.

7.2　MATHEMATICAL MODEL

Maxwell's equations:

For magnetic fields:

$$\left(\nabla^2 + k^2\right)\begin{pmatrix} H_x \\ H_y \\ H_z \end{pmatrix} = \begin{pmatrix} \left(\nabla \times J\right)_x \left(x, y, z, \omega\right) \\ \left(\nabla \times J\right)_y \left(x, y, z, \omega\right) \\ \left(\nabla \times J\right)_z \left(x, y, z, \omega\right) \end{pmatrix};$$

(7.1)

Similarly, for electric fields:

J is the current density

ψ_n is the function

a_n, b_n are the amplitude coefficients

$u_m n_P$, $v_m n_p$ are the Fourier basis function

hmn is the cut-off frequency

$$\begin{pmatrix} E_x \\ E_y \\ E_z \end{pmatrix} = \begin{bmatrix} \sum a_n \psi_n + \sum b_n \varphi_n \\ \sum a_n \psi'_n + \sum b_n \varphi'_n \\ \sum a_n \psi''_n + \sum b_n \varphi''_n \end{bmatrix}$$

$$= \sum a_n \begin{bmatrix} \psi_n \\ \psi'_n \\ \psi''_n \end{bmatrix} + \sum b_n \begin{bmatrix} \varphi_n \\ \varphi'_n \\ \varphi''_n \end{bmatrix}$$

$$\begin{pmatrix} E_x \\ E_y \\ E_z \end{pmatrix} = \frac{1}{h(mn)^2} \frac{\partial^2 u_{mnp}}{\partial z \partial x} C_{mnp};$$

(7.2)

$$\underline{E} = \begin{pmatrix} \sum_{mnp} \frac{C_{(mnp)}}{h(mn)^2} \frac{\partial^2 u_{mnp}}{\partial x \partial z} - \sum \frac{\mu d(mnp)}{h(mn)^2} j\omega(m,n,p) \frac{\partial^2 v_{mnp}}{\partial y \partial z} \\ \sum_{mnp} \frac{1}{h(mn)^2} C_{(mnp)} \frac{\partial^2 u_{mnp}}{\partial y \partial z} + \sum \frac{\mu}{h(mn)^2} j\omega(m,n,p) d_{(mnp)} \frac{\partial^2 v_{mnp}}{\partial x \partial z} \end{pmatrix} e^{j\omega(mnp)t};$$

(7.3)

$\underline{E}\left(x, y, z, t\right)$ = electric field component

$$= \sum_{m\,n\,p} C(mnp) \begin{pmatrix} \psi^E_{mnp_x}(x,y,z) \\ \psi^E_{mnp_y}(x,y,z) \\ \psi^E_{mnp_z}(x,y,z) \end{pmatrix} e^{j\omega(mnp)t}$$

$$+ \sum_{m\,n\,p} d(mnp) \begin{pmatrix} \varphi^E_{mnp_x}(x,y,z) \\ \varphi^E_{mnp_y}(x,y,z) \\ \varphi^E_{mnp_z}(x,y,z) \end{pmatrix} e^{j\omega(mnp)t}; \tag{7.4}$$

With duality:

$\underline{H}\ (x,y,z,t) =$ magnetic field component

$$= \sum_{m\,n\,p} C(mnp) \begin{pmatrix} \psi^H_{mnp_x}(x,y,z) \\ \psi^H_{mnp_y}(x,y,z) \\ \psi^H_{mnp_z}(x,y,z) \end{pmatrix} e^{j\omega(mnp)t} + \sum_{m\,n\,p} d(mnp) \begin{pmatrix} \varphi^H_{mnp_x}(x,y,z) \\ \varphi^H_{mnp_y}(x,y,z) \\ \varphi^H_{mnp_z}(x,y,z) \end{pmatrix} e^{j\omega(mnp)t}; \tag{7.5}$$

$\hat{n} \times H = J_s =$ surface current density on walls;

$$H_{Sx}(x,y,o) = J_{Sy}(x,y)$$

$$H_{Sy}(x,y,o) = -J_{Sx}(x,y)$$

Hence, $J_{Sy}(x,y,\delta t) =$ current density into RDRA

$$\sum_{mnp} C(mnp)\psi^H_{mnp_x}(x,y,o)e^{j\omega(mnp)t} + \sum_{mnp} d(mnp)\varphi^H_{mnp_x}(x,y,o)e^{j\omega(mnp)t}, \tag{7.6}$$

$$-J_{Sx}(x,y,dt) = \sum_{mnp} C(mnp)\psi^H_{mnp_y}(x,y,o)e^{j\omega(mnp)t}$$

$$+ \sum_{mnp} d(mnp)\varphi^H_{mnp_y}(x,y,o)e^{j\omega(mnp)t}; \tag{7.7}$$

If d is the length of probe inserted into DRA,

ψ^H, $\psi^E \varphi^H \varphi^E$ equations, we get from linear combinations

$$\psi^H_{mnp_x}(x,y,o) \propto \cos\cdot\sin \tag{7.8}$$

$$\psi^H_{mnp_y}(x,y,o) \propto \sin\cdot\cos \tag{7.9}$$

$$\varphi^H_{mnp_x}(x,y,o) \propto \sin\cdot\cos \tag{7.10}$$

$$\varphi^H_{mnp_y}(x,y,o) \propto \cos\cdot\sin \tag{7.11}$$

Amplitude coefficients can be determined from principle of orthonormality:

$C(mnp)\langle \psi_{mnp_y}^H (x,y,o), \psi_{mnp_y}^H \rangle$ Inner products or reaction terms can be written as:

$$C(mnp)\langle \varphi_{mnp_x}^H, \psi_{mnp_x}^H \rangle + d(mnp)\langle \varphi_{mnp_x}^H, \psi_{mnp_x}^H \rangle$$

$$= \lim_{T\to\infty} \frac{1}{2T} \int_{T}^{+T} < J_{Sy}(x,y,t), \psi_{mnp_x}^H (x,y,o) > e^{-j\omega(mnp)t} dt; \qquad (7.12)$$

and

$$C(mnp)\langle \varphi_{mnp_x}^H, \psi_{mnp_x}^H \rangle + d(mnp)\langle \varphi_{mnp_x}^H, \psi_{mnp_x}^H \rangle$$

$$= \lim_{T\to\infty} \frac{1}{2T} \int_{-T}^{+T} < J_{Sx}(x,y,t), \psi_{mnp_x}^H (x,y,o) > e^{-j\omega(mnp)t} dt. \qquad (7.13)$$

These are the solutions of amplitude coefficients $d(mnp)$ and $c(mnp)$ using the time averaging KAM method.

Following are the solutions of hybrid modes.

TE, TM, HEM modes $\to \omega(mnp)$ (Resonant mode).

$$\frac{\partial^2 E_z}{\partial x \partial z} = 0, \qquad z = o, d;$$

$$\frac{\partial E_z}{\partial z} = 0, \qquad z = 0, d;$$

For a homogeneous medium without source terms:

$$E_z = \Sigma C(mnp) e^{j\omega(mnp)t} u(mnp)(x,y,z,t)$$

$$H_z = \Sigma d(mnp) e^{j\omega(mnp)t} v(mnp)(x,y,z,t)$$

7.3   MODES IN HOMOGENOUS MEDIUM WITH SOURCE TERMS

For a homogeneous medium case:

$$E^{(hom)}(x,y,z,t) = \Sigma C(mnp) e^{j\omega(mnp)t} \psi_{mnp}^E(\underline{r}) + \Sigma d(mnp) e^{j\omega(mnp)t} \psi_{mnp}^E(\underline{r}); \qquad (7.14)$$

$r = x, y, z$

$$\underline{H}^{(hom)}\left(x,y,z,t\right)$$

$$= \Sigma C\left(mnp\right)e^{j\omega\left(mnp\right)t}\;\underline{\psi}_{mnp}^{H}\left(\underline{r}\right)\;+\Sigma d\left(mnp\right)e^{j\omega\left(mnp\right)t}\;\underline{\psi}_{mnp}^{H}\left(\underline{r}\right); \tag{7.15}$$

Hybrid modes can be generated by introducing non-resonant terms.

$$C_y^s\left(mnp\right)\quad d_y^s\left(mnp\right)$$

$$C_z^s\left(mnp\right)\quad d_z^s\left(mnp\right)$$

Thus, complete solution of hybrid modes has been obtained.

7.4   MATHEMATICAL MODELING OF HYBRID MODES

$$n\times H = J_s\;\;\text{on each wall}$$

$$n\times E = M_s\;\;\text{on each wall}$$

7.4.1   The Mathematical Derivation of Hybrid Modes Using Maxwell's Equation

First we develop the solution of a rectangular wave guide and switch to resonator. The wave guide solution is very simple. These wave guide equations will have both the fields H_z and E_z as given below:

$$\left(\nabla_\perp - \gamma\,\hat{z}\right)\times\left(E_z\,\hat{z}+\mathrm{E}_\perp\right)=-j\omega\mu\left(H_z\,\hat{z}+\mathrm{H}_\perp\right) \tag{7.16}$$

$$\left(\nabla_\perp - \gamma\,\hat{z}\right)\times\left(H_z\,\hat{z}+\mathrm{H}_\perp\right)=-j\omega\epsilon\left(E_z\,\hat{z}+\mathrm{E}_\perp\right) \tag{7.17}$$

$$\nabla_\perp E_z\times\hat{z}-\gamma\,\hat{z}\times\mathrm{E}_\perp =-j\omega\mu\mathrm{H}_\perp \tag{7.18}$$

$$\nabla_\perp H_z\times\hat{z}-\gamma\,\hat{z}\times\mathrm{H}_\perp =-j\omega\epsilon\mathrm{E}_\perp \tag{7.19}$$

$$\nabla_\perp E_z+\gamma\,\mathrm{E}_\perp =-j\omega\mu\,\hat{z}\times\mathrm{H}_\perp \tag{7.20}$$

$$\nabla_\perp E_z+\gamma\,\mathrm{E}_\perp =\frac{-j\omega\mu}{\gamma}\left(\nabla_\perp H_z\times\hat{z}-j\omega\epsilon\,\mathrm{E}_\perp\right) \tag{7.21}$$

$$\nabla_\perp E_z+\frac{j\omega\mu}{\gamma}\nabla_\perp H_z\times\hat{z}=-\left(\frac{\omega^2\mu\epsilon}{\gamma}+\gamma\right)\mathrm{E}_\perp$$

$$\gamma^2+\omega^2\mu\epsilon = h^2$$

Wave guide equations based on Helmholtz equations are:

$$E_\perp = \frac{-\gamma}{h^2} \nabla_\perp E_z - \frac{j\omega\mu}{h^2} \nabla_\perp H_z \times \hat{z} \qquad (7.22)$$

$$H_\perp = \frac{-\gamma}{h^2} \nabla_\perp H_z + \frac{j\omega\mu}{h^2} \nabla_\perp E_z \times \hat{z} \qquad (7.23)$$

Resonator equations simply replace $[\gamma]$ by $\left[-\dfrac{d}{dz} \right]$:

$$E_\perp = -\frac{1}{h^2} \frac{d}{dz} \nabla_\perp E_z - \frac{\mu}{h^2} \frac{d}{dt} \nabla_\perp H_z \times \hat{z} \qquad (7.24)$$

$$H_\perp = \frac{1}{h^2} \frac{d}{dz} \nabla_\perp H_z + \frac{\mu}{h^2} \frac{d}{dt} \nabla_\perp E_z \times \hat{z} \qquad (7.25)$$

$$\left(\nabla^2 + h^2 \right) \begin{pmatrix} E_z \\ H_z \end{pmatrix} = 0; \quad \text{Helmholtz equation}$$

Per the boundary conditions in DRA, when top and bottom walls of a resonator are PEC, the other four side walls are PMC.

$$H_z = 0; \quad \text{at } x = 0, a, \text{ and } y = 0, d; z = 0, d;$$

$$E_x = E_y = 0; \text{ at } z = 0, d;$$

$$H_x = 0; \quad \text{at } y = 0, b;$$

$$H_y = 0; \quad \text{at } x = 0, a;$$

Standard fields are:

$$H_z = \sum Re\left(d(mnp) e^{j\omega(mnp)t} v_{mnp}(\underline{r}) \right) \qquad (7.26)$$

$$E_z = \sum Re\left(c(mnp) e^{j\omega(mnp)t} u_{mnp}(\underline{r}) \right) \qquad (7.27)$$

Standard orthogonal fields are:

$$v_{mnp} = \frac{2\sqrt{2}}{\sqrt{abd}} \sin\left(\frac{m\pi x}{a} \right) \sin\left(\frac{m\pi y}{b} \right) \sin\left(\frac{m\pi z}{d} \right) \qquad (7.28)$$

$$u_{mnp} = \frac{2\sqrt{2}}{\sqrt{abd}} \cos\left(\frac{m\pi x}{a} \right) \cos\left(\frac{m\pi y}{b} \right) \cos\left(\frac{m\pi z}{d} \right) \qquad (7.29)$$

The above equations have been obtained from expansion of Helmholtz equation by the separation of variables method:

$$\left(\nabla^2 + h^2\right)H_z = 0$$

Hence, $h^2 = h_{mn}^2 = \pi^2\left(\frac{m^2}{a^2} + \frac{n^2}{b^2}\right)$; this gives the resonant frequency of RDRA. The tensor product of linear combination can appear as given below:

$$H_z = \mathcal{L}\left\{\begin{array}{l} \cos\left(\dfrac{m\pi x}{a}\right)\cos\left(\dfrac{n\pi x}{a}\right), \ \cos\left(\dfrac{n\pi x}{a}\right)\sin\left(\dfrac{m\pi x}{a}\right), \\[3mm] \sin\left(\dfrac{m\pi x}{a}\right)\cos\left(\dfrac{n\pi x}{a}\right), \ \sin\left(\dfrac{n\pi x}{a}\right)\sin\left(\dfrac{m\pi x}{a}\right) \end{array}\right\}$$

where $\mathcal{L}$ denotes linear combinations. It turns out that, depending on the nature of wall or surface (PEC or PMC), four possible linear combinations can appear $\left(\cos\otimes\sin, \sin\otimes\cos, \text{and } \sin\otimes\sin, \cos\otimes\cos\right)$.

Also, $\omega^2\mu\epsilon + \gamma^2 = h_{mn}^2$

Hence,

$H_z = 0$ when $x = 0$, cos terms are ruled out from x.

$H_z = 0$ when $y = 0$, again cos terms are ruled out from y.

$$H_z = \sin\left(\frac{m\pi x}{a}\right)\sin\left(\frac{n\pi y}{b}\right)\left(C_1 e^{\gamma_{mnz}} + C_2 e^{-\gamma_{mnz}}\right)$$

$$H_z = 0 \ ; \quad \text{when} \quad z = 0, d$$

$$(C_1 + C_2) = 0$$

$$\left(e^{\gamma_{mnd}} - e^{-\gamma_{mnd}}\right) = 0, \beta$$

$$C_1 = C_2 \ ; \quad \sin\left(\gamma_{mnd}\right) = 0$$

$$\gamma_{mn} = j\beta_{mn}$$

$$\beta_{mnd} = \pi p$$

Hence, $\beta_{mn} = \dfrac{p\pi}{d}$

$$H_z = \sin\left(\frac{m\pi x}{a}\right)\sin\left(\frac{n\pi y}{b}\right)\sin\left(\frac{p\pi z}{d}\right) \tag{7.30}$$

$\omega^2\mu\epsilon - \left(\frac{p\pi}{d}\right)^2 = h_{mn}^2$; hence, resonant frequency can be determined as

$\omega\sqrt{\mu\epsilon} = \pi\sqrt{\frac{m^2}{a^2} + \frac{n^2}{b^2} + \frac{p^2}{d^2}}$. Here, we note that resonant frequency in hybrid mode is the same for TE and TM modes.

Now

$$H_x = \frac{1}{h^2}\frac{\partial}{\partial z}\frac{\partial H_z}{\partial x} + \frac{\mu}{h^2}\frac{\partial}{\partial t}\frac{\partial E_z}{\partial y};$$

$$H_x = 0 \quad \text{at } y = 0, b;$$

$$\frac{\delta H_x}{\delta x} = 0 \quad \text{at } y = 0, b;$$

$$\frac{\delta E_z}{\delta y} = 0 \; ; \quad \text{at } y = 0, b;$$

$$E_x = 0;$$

$$E_y = 0, \; z = 0, d;$$

Hence, $E_x = \dfrac{1}{h^2}\dfrac{\partial}{\partial z}\dfrac{\partial}{\partial x}E_z - \dfrac{\mu}{h^2}\dfrac{\partial}{\partial t}\dfrac{\partial H_z}{\partial y};$

$$\frac{\partial^2 E_z}{\partial x \partial z} = 0 \quad \text{at } z = 0, d$$

$$\frac{\partial E_z}{\partial t} = 0 \quad \text{at } z = d$$

z-dependence of E_z is $\cos\left(\dfrac{\pi p z}{d}\right)$, $E_x = 0$; when $x = 0, a$;

$$E_z = \cos\left(\frac{m\pi x}{a}\right)\cos\left(\frac{n\pi y}{b}\right)\cos\left(\frac{p\pi z}{d}\right)$$

This is the way of getting E_z and H_z longitudinal components by method of separation of variables.

7.5   GENERAL SOLUTION OF HYBRID MODES (HEM)

The investigations are based on first applying waveguide theory, which models the electromagnetic fields to vary with z-axis, i.e., these are exploited into the Maxwell curl equations, then manipulating them to express the transverse components of the fields in terms of partial derivatives of the longitudinal components of the fields w. r. t. x and y axis (i.e., the transverse coordinates). Wave guide models of four different rectangular DRAs with specified boundary conditions filled with homogeneous material having linear permittivity have been mathematically developed and realized to determine TE and TM modes propagating fields. These have resulted into different sine–cosines combinations. Propagation of these fields has been split as inside the DRA and outside with an interfacing surface having two different permittivity on both sides. The solution is developed as a transcendental equation which purely characterizes rectangular DRA resonant frequency and propagating fields. The amplitude coefficient of these fields $Cmnp$ and $Dmnp$ inside the DRA can be determined by comparing time averaged magnetic energies equal to time averaged electrical energies by the KAM method based on the principle of orthonormality. The transverse components of E_x, E_y, H_x, H_y are the components determined in terms of longitudinal components E_z, H_z. These transverse fields satisfy Helmholtz equations, are expressible in terms of $u(mnp) e^{j\omega(mnp)t}$ and $v(mnp) e^{j\omega(mnp)t}$. E_x, E_y, H_x, H_y fields are also expressible in terms of partial derivatives of $u(mnp)$, $v(mnp)$. Hence if $C(mnp)$, $D(mnp)$ denotes the linear combinational coefficients of $u(mnp)$, $v(mnp)$ for E_z, H_z, then the same coefficients appear in E_x, E_y, H_x, H_y. These coefficients (mnp), $D(mnp)$ can be determined by matching H_x, H_y at $z = 0$ to the surface current density of RDRA, when feed at $z = 0$. If the surface excitation at $z = 0$ has a frequency component other than $\omega(mnp)$, say ω, then the fields amplitude components corresponding to this excitation are determined. Both the fields E_z and H_z will remain excited at any instant of time in resonator, then these modes can be termed as hybrid modes. Our solution is developed based on homogeneous medium in the resonator.

Hence, general hybrid equations can be written as follows:

$$\underline{E} = \sum \left[c_{(mnp)} \, \underline{\Psi}^E_{mnp}\left(\underline{r}\right) + d\left(mnp\right)\underline{\phi}^E_{mnp}\left(\underline{r}\right) \right] e^{-j\omega(mnp)t}$$

$$\underline{H} = \sum \left[c_{(mnp)} \, \underline{\Psi}^H_{mnp}\left(\underline{r}\right) + d\left(mnp\right)\underline{\phi}^H_{mnp}\left(\underline{r}\right) \right] e^{-j\omega(mnp)t}$$

A solution of the DRA can be developed by using these above two equations. For this, we insert a probe of δ length having R radius into rectangular DRA. This is pointing toward the z-axis.

$$x = \frac{a}{2} + R\cos\phi;$$

$$y = \frac{b}{2} + R\sin\phi;$$

$$z = 0;$$

expressed based on Cartesian to cylindrical coordinates.

$\hat{n} \times \underline{H} = \underline{J}_s$: This is based on boundary conditions inside the RDRA.

$$\hat{\rho}\left(H_z\hat{z} + H_\phi\hat{\phi}\right) = \underline{J}_s$$

$$H_\phi\hat{z} - H_z\hat{\phi} = \underline{J}_s$$

and

$$H_\phi\left(\frac{a}{2} + R\cos\phi, \frac{b}{2} + R\sin\phi, z\right) = J_{sz}\left(\phi, z\right) \qquad \left(0 < z < \delta, \;\; 0 < \phi < 2\pi\right)$$

$$H_z\left(\frac{a}{2} + R\cos\phi, \frac{b}{2} + R\sin\phi, z\right) = J_{s\phi}\left(\phi, z\right)$$

$$-J_{s\phi}\left(\phi, z, t\right) \;\;\; = \Sigma d\left(mnp\right)v_{mnp}\left(\frac{a}{2} + R\cos\phi, \frac{b}{2} + R\sin\phi, z\right)e^{j\omega\left(mnp\right)t}$$

$$H_\phi\left(\frac{a}{2} + R\cos\phi, \frac{b}{2} + R\sin\phi, z\right) = -H_x\sin\phi + H_y\cos\phi$$

$$= J_{sz}\left(\phi, z\right) \qquad \left(0 < z < \delta, \;\; 0 < \phi < 2\pi\right)$$

$$= -J_{s\phi}\left(\phi, z\right)$$

$$J_{sz}\left(\phi, z, t\right) = \sin\phi\sum\begin{bmatrix} -c_{(mnp)}\dfrac{j\mu\omega\left(mnp\right)}{h_{(mn)}^2}\dfrac{\partial v_{mnp}}{\partial y}\left(\dfrac{a}{2} + R\cos\phi, \dfrac{b}{2} + R\sin\phi, z\right) \\[4mm] -\dfrac{d_{(mnp)}}{h_{(mn)}^2}\dfrac{\partial^2 v_{mnp}}{\partial z\,\partial x}\left(\dfrac{a}{2} + R\cos\phi, \dfrac{b}{2} + R\sin\phi, z\right) \end{bmatrix}$$

$$e^{+j\omega\left(mnp\right)t} + \cos\phi\sum\begin{bmatrix} c_{(mnp)}\left(-\dfrac{j\mu\omega\left(mnp\right)}{h_{(mn)}^2}\right)\dfrac{\partial}{\partial x}u_{mnp}\left(\dfrac{a}{2} + R\cos\phi, \dfrac{b}{2} + R\sin\phi, z\right) \\[4mm] +\dfrac{d_{(mnp)}}{h_{(mn)}^2}\dfrac{\partial^2 v_{mnp}}{\partial z\,\partial x}\left(\dfrac{a}{2} + R\cos\phi, \dfrac{b}{2} + R\sin\phi, z\right) \end{bmatrix}$$

$$e^{+j\omega\left(mnp\right)t};$$

$$-J_{s\phi}(\phi,z,t) = \sum_{mnp} c_{(mnp)} e^{j\omega(mnp)t} X_{mnp}(\phi,z);$$

where $X_{mnp}(\phi,z) = v_{mnp}\left(\dfrac{a}{2} + R\cos\phi, \dfrac{b}{2} + R\sin\phi, z\right);$

$$J_{sz}(\phi,z,t) = \sum_{mnp} c_{(mnp)} \eta_{mnp}^{(1)}(\phi,z) + d_{(mnp)} \eta_{mnp}^{(2)}(\phi,z) e^{j\omega(mnp)t}$$

where $\eta_{mnp}^{(1)}(\phi,z) = \dfrac{-j\omega(mnp)t\,\mu\sin\phi}{h_{(mn)}^2}\dfrac{\partial}{\partial y} u_{mnp}\left(\dfrac{a}{2} + R\cos\phi, \dfrac{b}{2} + R\sin\phi,\ z\right)$

$$\eta_{mnp}^{(2)}(\phi,z) = \dfrac{-\sin\phi}{h_{(mn)}^2}\dfrac{\partial^2 v_{mnp}}{\partial x\partial z}\left(\dfrac{a}{2} + R\cos\phi, \dfrac{b}{2} + R\sin\phi, z\right) e^{-j\omega(mnp)t}\ \tilde{X}_{mnp}(\phi,z)$$

$$C_{(mnp)} = \lim_{T\to\infty} -\frac{1}{2T}\int J_{s}\phi(\phi,z,t)\tilde{X}_{mnp}(\phi,z)e^{-j\omega(mnp)t}\,dt\,d\phi\,dz$$

$$|t| < T$$

$$0 < \phi < 2\pi$$

$$0 < z < \delta$$

$$\int \left|\tilde{X}_{mnp}(\phi,z)\right|^2 d\phi\,dz$$

$$c_{(mnp)}\int \left[\left|\eta_{mnp}^{(1)}(\phi,z)\right|^2 d\phi dz + d_{(mnp)}\int \overline{\eta_{mnp}^{(1)}(\phi,z)}\eta_{(mnp)}^{(2)}(\phi,z)\right]d\phi\,dz$$

$$= \lim_{T\to\infty}\frac{1}{2T}\int \overline{\eta_{mnp}^{(1)}(\phi,z)}\,J_{sz}(\phi,t)e^{-j\omega(mnp)t}\ d\phi\,dz\,dt$$

$$|t| \le T$$

$$c_{(mnp)} \int \eta_{mnp}^{(1)}(\phi,z) \eta_{mnp}^{(2)}(\phi,z)\, d\phi\, dz + d_{(mnp)} \int \left| \eta_{mnp}^{(2)}(\phi,z) \right|^2 d\phi\, dz$$

$$= \lim_{T \to \infty} \frac{1}{2T} \int J_{sz}(\phi,z,t)\, \eta_{mnp}^{(2)}(\phi,z) e^{-j\omega(mnp)t}\, d\phi\, dz\, dt \qquad (7.31)$$

$$|t| < T$$

If we keep $J_{s\phi} = 0$, from Equations (7.23) and (7.24), we get $C(mnp)$ and $d(mnp)$.

7.6 CONCLUSION

Hybrid modes can introduce diversity into reception and transmission in communication systems. They can also provide frequency conservation. One frequency can transmit two signals at the same time. Similar is the case for reception on using hybrid modes. In sapphire crystals, formation of multi beam reflection is easy using hybrid modes. Gems have a crystalline nature and can generate hybrid modes. Excitation of hybrid mode in DRA is a complex task. Short and open boundaries are the basis of modes. The half wavelength resonant modes with odd numbers only will be excited when ground plane is used.

Even modes get short circuited due to ground plane. The higher modes will have higher resonant frequency. The number of higher modes also modifies the radiation patterns, i.e., the mode number will be equal to the number of lobes in the final radiation pattern. Care must be taken to select this hybrid number n because it has a direct relationship with the radiation pattern of far fields or beam shape. Gain of antenna can also decrease abruptly due to dispersion at higher modes. This is introduced when the dipole moment starts overlapping. Based on various solutions, hybrid modes can be memorized for any particular mode with desired radiation patterns. The polarization of even and odd modes is opposite. Highly directive patterns can be obtained at higher modes. Bandwidth of higher order modes will be decreased. HEM 1, 3, 5, 7,... are odd modes that can be written as HE. Similarly HEM 2, 4, 6, 8,... are even modes or EH mode. In hybrid modes E_z as well as H_z, fields can propagate at the same time. The HEM are also HE (odd hybrid modes) and EH (even hybrid modes). They can impart wide design space in the field of antenna. These designs can be used in beam control and regulation.

REFERENCES

1. R. K. Mongia and P. Bhartia, "Dielectric resonator antennas—a review and general design relations for resonant frequency and bandwidth," *Int. J. Microw. Millimeter-Wave Comput. Eng.*, vol. 4, no. 3, p. 230–247, 1994.
2. R. K. Mongia and A. Ittipiboon, "Theoretical and Experimental Investigations on Rectangular Dielectric Resonator Antennas," *IEEE Trans. Antennas Propag.*, vol. 45, no. 9, pp. 1348–1356, 1997.

3. B. Y. Toh, R. Cahill, and V. F. Fusco, "Understanding and Measuring Circular Polarization," *IEEE Trans. Educ.*, vol. 46, no. 3, pp. 313–318, 2003.

4. M. B. Oliver, R. K. Mongia, and Y. M. M. Antar, "A New Broadband Circularly Polarized Dielectric Resonator Antenna," *IEEE Antennas Propag. Soc. Int. Symp. 1995 Dig.*, vol. 1, pp. 4–7, 1995.

5. C. Huang, J. Wu, and K. Wong, "Cross-slot-Coupled Microstrip Antenna and Dielectric Resonator Antenna for Circular Polarization," *IEEE Trans. Antennas Propag.*, vol. 47, no. 4, pp. 605–609, 1999.

6. G. Almpanis, C. Fumeaux, and R. Vahldieck, "Offset Cross-slot-Coupled Dielectric Resonator Antenna for Circular Polarization," *IEEE Microw. Wirel. Components Lett.*, vol. 16, no. 8, pp. 461–463, 2006.

7. X. S. Fang and K. W. Leung, "Linear-/Circular-Polarization Designs of Dual-/Wide-Band Cylindrical Dielectric Resonator Antennas," *IEEE Trans. Antennas Propag.*, vol. 60, no. 6, p. 2662–2671, 2012.

8. Y. Pan, K. W. Leung, and E. H. Lim, "Compact Wideband Circularly Polarised Rectangular Dielectric Resonator Antenna with Dual Underlaid Hybrid Couplers," *Microw. Opt. Technol. Lett.*, vol. 52, no. 12, pp. 2789–2791, 2010.

9. R. Chair, S. L. S. Yang, A. A. Kishk, K. F. Lee, and K. M. Luk, "Aperture Fed Wideband Circularly Polarized Rectangular Stair Shaped Dielectric Resonator Antenna," *IEEE Trans. Antennas Propag.*, vol. 54, no. 4, pp. 1350–1352, 2006.

10. K. X. Wang and H. Wong, "A Circularly Polarized Antenna by Using Rotated-Stair Dielectric Resonator," *IEEE Antennas Wirel. Propag. Lett.*, vol. 14, pp. 787–790, 2015.

11. Y. M. Pan and K. W. Leung, "Wideband Omnidirectional Circularly Polarized Dielectric Resonator Antenna with Parasitic Strips," *IEEE Trans. Antennas Propag.*, vol. 60, no. 6, pp. 2992–2997, 2012.

12. Y. Pan and K. W. Leung, "Wideband Circularly Polarized Trapezoidal Dielectric Resonator Antenna," *IEEE Antennas Wirel. Propag. Lett.*, vol. 9, pp. 588–591, 2010.

13. M. Khalily, M. R. Kamarudin, M. Mokayef, and M. H. Jamaluddin, "Omnidirectional Circularly Polarized Dielectric Resonator Antenna for 5.2-GHz WLAN Applications," *IEEE Antennas Wirel. Propag. Lett.*, vol. 13, pp. 443–446, 2014.

14. G. Varshney, V. S. Pandey, R. S. Yaduvanshi, and L. Kumar, "Wide Band Circularly Polarized Dielectric Resonator Antenna with Stair-Shaped Slot Excitation," *IEEE Trans. Antennas Propag.*, vol. 65, no. 3, p. 1380–1383, March, 2017.

15. G. Varshney, V. S. Pandey, and R. S. Yaduvanshi, "Dual-Band Fan-Blade-Shaped Circularly Polarised Dielectric Resonator Antenna," *IET Microwaves, Antennas Propag.*, vol. 11, no. 13, pp. 1868–1871, 2017.

16. G. Varshney, S. Gotra, V. S. Pandey, and R. S. Yaduvanshi, "Inverted-Sigmoid Shaped Multiband Dielectric Resonator Antenna with Dual-Band Circular Polarization," *IEEE Trans. Antennas Propagat.*, vol. 66, no. 4, pp. 2067–2072, 2018.

17. M. Khalily, M. R. Kamarudin, and M. H. Jamaluddin, "A Novel Square Dielectric Resonator Antenna with Two Unequal Inclined Slits for Wideband Circular Polarization," *IEEE Antennas Wirel. Propag. Lett.*, vol. 12, pp. 1256–1259, 2013.

18. J. Pan and M. Zou, "Wideband Hybrid Circularly Polarised Rectangular Dielectric Resonator Antenna Excited by Modified Cross-slot," *Electron. Lett.*, vol. 50, no. 16, pp. 1123–1125, 2014.

19. M. Zou, J. Pan, and Z. Nie, "A Wideband Circularly Polarized Rectangular Dielectric Resonator Antenna Excited by an Archimedean Spiral Slot," *IEEE Antennas Wirel. Propag. Lett.*, vol. 14, pp. 446–449, 2015.

20. G. Varshney, V. S. Pandey, and R. S. Yaduvanshi, "Axial Ratio Bandwidth Enhancement of a Circularly Polarized Rectangular Dielectric Resonator Antenna Gaurav Varshney, V. S. Pandey and R. S. Yaduvanshi," *Int. J. Microw. Wirel. Technol. Cambridge Univ. Press*, vol. (accepted), 2018.

21. R. K. Mongia, "Theoretical and Experimental Resonant Frequencies of Rectangular Dielectric Resonators," *IEE Proc. H Microwaves, Antennas Propag.*, vol. 139, no. 1, p. 98, 1992.

22. T. Chang and J. Kiang, "Bandwidth Broadening of Dielectric Resonator Antenna by Merging Adjacent Bands," *IEEE Trans. Antennas Propag.*, vol. 57, no. 10, pp. 3316–3320, 2009.

23. A. Buerkle, K. Sarabandi, and H. Mosallaei, "Compact Slot and Dielectric Resonator Antenna with Dual-Resonance, Broadband Characteristics," *IEEE Trans. Antennas Propag.*, vol. 53, no. 3, pp. 1020–1027, 2005.

24. A. Petosa, *Dielectric Resonator Antenna Handbook*. London, UK: Artech House, 2007.

25. K. M. Luk and K. W. Leung, *Dielectric Resonator Antennas*. Baldock, UK: Research Studies Press Ltd., 2003.

26. S. Fakhte, H. Oraizi, R. Karimian, and R. Fakhte, "A New Wideband Circularly Polarized Stair-Shaped Dielectric Resonator Antenna," *IEEE Trans. Antennas Propag.*, vol. 63, no. 4, pp. 1828–1832, 2015.

8 Nano Dielectric Resonator Antenna (NDRA)

8.1 INTRODUCTION

Nano DRA is a terahertz antenna which can be used for medical sensing, imaging, energy harvesting, and ultrafast data transfer due to ultra-large bandwidth. Terahertz (THz) waves are absolutely harmless to humans and have no ionizing radiation, unlike X-ray machines. The source of coherent pulsed THz radiation is photoconductive antennas. The rapid variations of surface photoconductivity of graphene and silicon dioxide substrates with optical excitation (Gaussian source) using LASER are achieved. Gallium arsenide (GaAs) is a second harmonic generation (SHG) material, and graphene with SiO_2 substrate is a third harmonic generation (THG) material. After exciting the gap between the two electrodes and use of LASER input excitations, the concentration of charge carriers increases into this region or gap. Being that graphene and silicon dioxide are nonlinear substrates of third harmonic generation, THz pulse generation occurs. The laser excitation is used to exploit several nonlinear optical effects that can provide high-contrast imaging of biological samples. The transmitted or reflected portion of each pulse is then detected after a time delay. Graphene is inhomogeneous material and is produced multimode. Terahertz light is dominated by the dielectric response of the materials with different absorption features compared with healthy tissue. Due to merging of higher order modes with fundamental mode at higher frequency, the bandwidth is increased. A sequence of microwave pulses is injected into a particular point on an object. The electromagnetic imaging may be perceived as the retrieval of the spatial distribution of the permittivity function of the medium. The terahertz's wavelength is located between the microwave and the infrared region of the electromagnetic spectrum. Because tissues constitute different molecules which varying water content, hence varying permitivity, thus based on permitivity measurements information is obtained through THz based scaning. Permittivity characteristics of THz waves on healthy and unhealthy tissues, their interaction with tissues is studied using an algorithm. THz capability based on the high concentration of the electric field is represented by its potential for sensing the complex dielectric properties of small compounds of a different nature. This limits our ability to design and manufacture NDRA.

Recent 5G technology requires large bandwidth. The requirements for India itself will be 9.8 GB of data per month for every smart phone user. There are 1.16 billion mobile users in India (data based on TRAI reports, April 2019). Therefore, a large

amount of bandwidth will be required for 5G networks. Patch and dielectric resonator antennas are popular at microwave frequencies with limited bandwidth for
wireless communication networks [1]. Accurate solutions for both the above types
of antennas are available in literature and in theoretical and experimental work.
Both types of research articles are available in mass with substantial results [2].
Theoretical physics on quantum field theory has been very interesting, and China
has gone ahead in this field. They have reported in the media development of quantum radars and quantum satellites. Quantum entanglement is an important phenomenon [3]. Much less literature is available worldwide in the field of quantum
antennas or NDRA [4]. It is evident that metal offers very high losses at optical
frequencies as compared to dielectric material [5]. Surface plasmonic resonance
occurs at interface of metal and substrate at optical frequencies [6]. NDRA makes
use of SPP phenomenon. Nanostrip waveguide is a popular feed in light antennas
at optical frequencies in contrast to microstrip/coaxial feed used at microwave frequencies [7]. A Gaussian beam is inserted into nano wave guide in place of lumped
port or wave port in optical modeling for NDRA [8]. The Schrodinger equation can
provide a solution for plasmonic resonance frequency and scattering of these
nonlinear parameters [9]. All transport parameters have been workedout based
on Drude nonlinear modeling [10–12]. SPP is used to drive nano DRA through
capacitive coupling known as proximity feed. This introduces radiation into space
with ceramic nano DRA [13–16]. Nanostrip waveguide acts as a resonator and is
used as an efficient coupler to DR for excitation and loading. In a silver nanostrip
waveguide, the incident wave does not reflect back completely at optical frequency
due to the Drude model. The radiation penetrates through metal and gives rise to
excitations of free elections. This phenomenon is also known as plasmon resonance
or SPP (surface plasmon positrons) [13,14]. The SPP is collective oscillation of free
electrons in noble metal nanostructures due to strong interaction of conduction
electrons with incident radiations. At resonant frequency, they result in scattering,
absorption, and local field enhancement [14,15]. This phenomenon is also named
creation or annihilation of photons, i.e., decay of plasmons into photons by means
of emission of photons or loss due to electron-hole excitation. Optical nano antennas, analogous to radio and microwave antennas, are devices that receive and emit
radiation in the visible region of the electromagnetic spectrum, corresponding to
frequencies of several 100 THz. The optical devices can directly convert an electric
current into visible light.

Higher order modes are generated by NDRA (nano dielectric resonator antenna).
Control on higher order modes has a significant impact on mode merging, mode
shifting, beam width control, gain control, and bandwidth control into NDRA [1].
NDRAs, being compact in size, can be used in arrays for specific applications.
They are the best candidates for futuristic wireless energy harvesting antennas [15].
They can be integrated with vehicles for energy resources directly from radars or
satellites available in environments around the clock. There is no requirement of sun
visibility in contrast to photovoltaic cells.

The Drude model has provided solutions to characterize metals when switching form microwave to optical frequencies. Silver nanostrip waveguide is used in SiO_2 substrate to couple laser beams by selecting Gaussian beam as excitation to NDRA. The primary challenge is to identify a suitable driving source which can be integrated with optical antennas at the nanoscale [16]. Here, we develop quantum interface for the conversion of electrical energy to photons, mediated by optical antennas.

In 1939, Richtmyer found that dielectric objects can resonate and radiate into free space based on boundary conditions like the air-dielectric interface.

In 1985, SA long was experimentally proved. Hence, DR are very efficient radiators.

DR material is commercially available with a vast range of permittivity ranging from 10 to 1600 farad/meter.

An antenna designer can exercise choice for exciting various higher order modes depending on material permittivity and dimensions of DR at optical frequency [17].

$$I(\lambda) = I_{inc}(\lambda) - I_{abs}(\lambda) - I_s(\lambda)$$

At optical frequency, shorter effective waveguide is used as compared to microwave regime.

The optical spectrum is 10^{13} to 10^{15} Hz. Optical antenna can couple optical energy into plasmonic resonance and vice versa. Plasmon resonance thus becomes the main cause of excitation to nano dielectric resonator antenna [18].

The quantum antenna consists of electrons, and positrons, are solved by Dirac second quantized field equations, these are based on quantum electrodynamics which described spin of photons. Here, the electromagnetic field is produced by linear superposition of creation and annihilation operators of photon fields. The electron positron creation and annihilation operator fields plus photons creation and annihilation operators' fields is the total field produced in quantum antenna [19]. The current density of fields is obtained by quadratic functions of the Dirac field operators. The current density produced a quantum electromagnetic field described by the retardation potentials as a Dirac wave function. Hence computations of mean and mean square fluctuations of quantum e.m. fields produced by antenna plus free photon e.m. fields in any state shall provide the complete solution. Dirac current moments are a combined state of electrons positrons and photon fields. Interaction of free photon fields and radiated fields of electrons positrons takes place to introduce spin fields known as current moments. Drude's modeling can provide a solution at optical frequencies [20–22]. Hence, in NDRA, electrons and positrons based on Dirac second quantized field equations develop radiations into space [17–24]. Current density thus produced in a quantum electromagnetic field is described by retarded potentials. Figures 8.1 through 8.4 are NDRAs. Figure 8.4b–d is excitation, E and H fields, respectively. Figures 8.20 and 8.21 are E-fields and dipoles. In this chapter at many places NDRA has been synonymously used as quantum antenna.

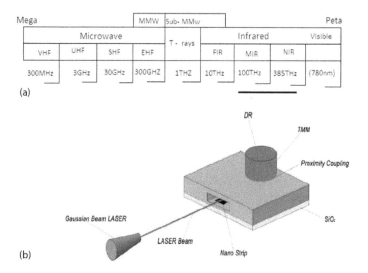

FIGURE 8.1 (a) Frequency spectrum showing microwave and optical frequency band and (b) design structure of nano DRA.

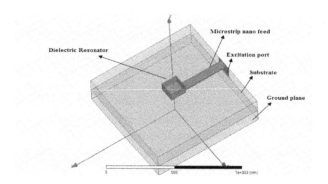

FIGURE 8.2 Rectangular NDRA with Gaussian input.

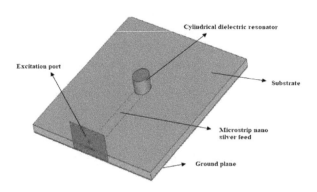

FIGURE 8.3 Cylindrical NDRA with Gaussian input.

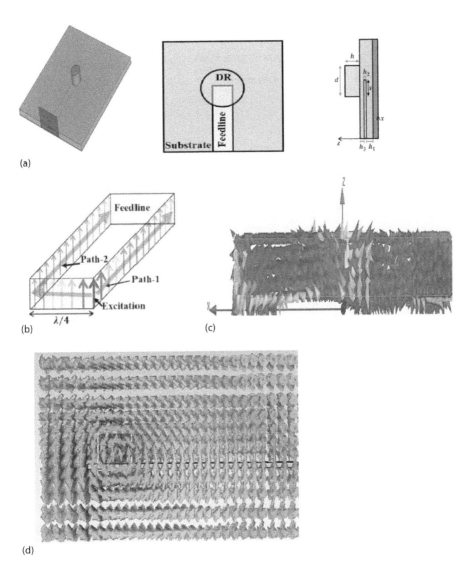

(a)

(b)

(c)

(d)

FIGURE 8.4 (a) CNDRA with radiations at 192 THz and Gaussian input at 483 THz to silver nano waveguide, (b) excitation field in nano waveguide, (c) E-field, and (d) H-fields.

8.2 DESIGN DESCRIPTION WITH RADIATED FIELDS OF NDRA

A Gaussian beam is used as input excitation to a silver nano waveguide. The silver nano waveguide is inserted into SiO_2 substrate. These laser pulses are expressed as femto/second. These laser pulses have higher photon energy than the band gap energy of a substrate. Hence, SPP is created at terahertz frequency. SPP derive DRA to introduce electromagnetic radiation. Dielectric resonators (CNDRA/RNDRA) of different geometry models for quantum antennas have been proposed. These designs are

TABLE 8.1
Dimensions of Quantum CDRA

S. No.	Parameter	Value
1	Radius "r"	0.250 µm
2	Height "h"	0.325 µm
3	Permittivity	11.9 f/m
4	Input frequency	473 THz
5	Radiated frequency	192 THz
6	Gain	10.5 dB
7	Substrate	Arlon
8	Dimension of the substrate	4.6 × 4.5 µm

TABLE 8.2
Dimension Table of QRDRA

S. No.	Name	Material Used	Permittivity	Length (nm)	Width (nm)	Height (nm)
1	Ground plane	Silver	1	900	900	100
2	Substrate	Teflon-based	2.08	900	900	100
3	Microstrip nano feed	Silver	1	450	67	10
4	Rectangular DR	Titanium dioxide	8.29	128	128	60
5	Vacuum sheet	Vacuum	1	1500	1500	1200

illustrated in Figures 8.1 and 8.2. Their design specifications are given in Tables 8.1 and 8.2. Drude scattering is function of frequency, resistance, and conductance. The quantum Boltzmann transport equation is used to provide an exact solution of transport parameters. Higher order modes can be excited into quantum antenna. Knowledge on resonant modes, as shown in Figure 8.21, gives physical insight of antennas during the design phase itself. Mode control in NDRA or quantum antenna can further extend the scope of applications such as mode merging, mode shifting, and generation of higher order modes.

In this antenna, photon generation is known as Bosonic fields or fermionic fields. This nonlinearity of photons is described by Dirac's second quantized field equations. Feynman path integral gives solution to photons moment fluctuations. Correlation coefficient manipulation provided proper beam formation. Photon spin is the main cause of nonlinearity in quantum antennas.

Two different models of quantum DRAs having different geometries have been designed and simulated for possible results based on theoretical models. Hardware of these models could not be developed because of non-availability of test facilities at present. However, these models have been simulated using CST 17 version software.

These simulated results have opened up a new field of research in the field of quantum antennas or NDRA with excellent simulated results at terahertz frequencies. Models have been shown with their diagrams and results obtained on RF simulators, i.e., HFSS or CST.

The quantum RDRA is excited by plasmonic resonance. Optical input is given into the SiO_2 substrate thorough a silver nano waveguide. An energy band gap of semiconductor material is targeted to accelerate by input excitation. Transient photo currents are thus produced. Accelerated charge carriers are thus generated. Therefore, oscillations in QRDRA, also known as resonance, are created at THz frequency. This results in radiation into space depending on boundary conditions of QRDRA. This radiation is governed by the rule of plasmonics. Using higher order modes concept dimension scaling can become possible. The quantum Boltzmann transport equation can provide an accurate solution to quantum antenna. The frequency of this antenna can be one terahertz to seven hundred terahertz.

This device can convert optical radiation into localized energy and vice versa. A proximity feed is used to excite QDRA. Poynting theorem describes power dissipated by a time harmonic system. A nano waveguide (silver) is inserted into substrate to create plasmonic resonance. Then through capacitive coupling, the near field is coupled to DRA for generating far field radiation. Quantum antenna is shown in Figure 8.4a and excitation and fields are presented in Figure 8.4b–d along with dimensions in Table 8.1. The working principle is described as below:

Main features of NDRA:

1. Light in light out.
2. Input excitation by semiconductor LASER.
3. Plasmonic resonance takes place in substrate (SPP).
4. Electromagnetic coupling takes place in QRDRA and QCDRA through plasmonic resonance.
5. Capacitive coupling takes place due to proximity feed. It forms photon clouds and fluctuations due to spin effect of photons.
6. Screening of electronic—hole takes place near band gap due to accelerated charge carriers.
7. Initially very small band gap energy is observed.
8. LASER provides higher photon energy then initial band gap energy.
9. LASER accelerates photons due to excitation known as plasmonic resonance.
10. LASER creates clouds of charge carrier (screening of fields takes place).
11. Quick change in field takes place thus generating transient currents.

Resonant frequency formulation of CNDRA (here, d and h are radius and height of CNDRA):

$$f = \frac{6.324c}{2\pi d \sqrt{\varepsilon_r + 2}} \left[0.27 + 0.36\left(\frac{d}{2h}\right) + 0.002\left(\frac{d}{2h}\right)^2 \right]$$

The penetration of EM fields into the metal ground plane takes place. Election oscillations cause skin effects. Drude scattering is a function of frequency, resistance, and conductance. QRDRA is an efficient radiator. It can excite an infinite number of resonant modes. Knowledge of resonant modes can predict antenna behavior. They can estimate characteristics of field produced. This can be verified using Gaussian beam excitation as per Figure 8.7 using CST software. Figures 8.5 through 8.11 present

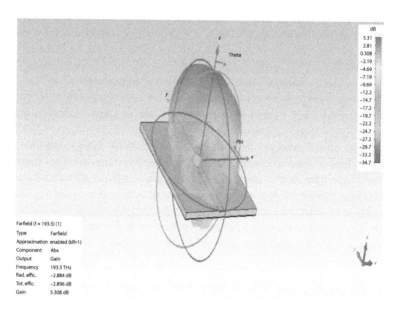

FIGURE 8.5 Quantum antenna radiation in CNDRA excited with wave port.

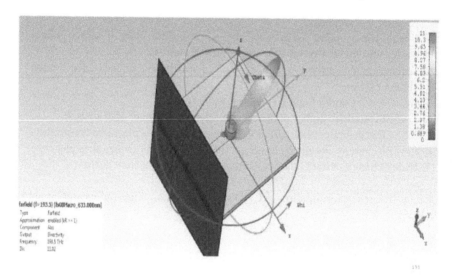

FIGURE 8.6 Radiation pattern of CNDRA with Gaussian beam input.

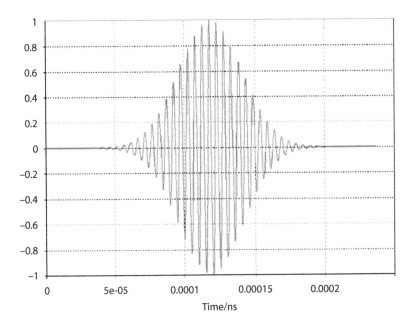

FIGURE 8.7 Gaussian input at 473 GHz input waveguide.

NANO DRA

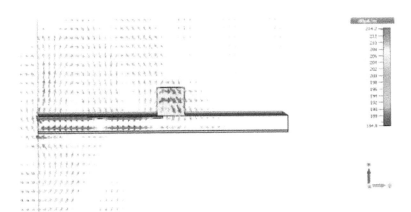

FIGURE 8.8 E-fields of nano CNDRA.

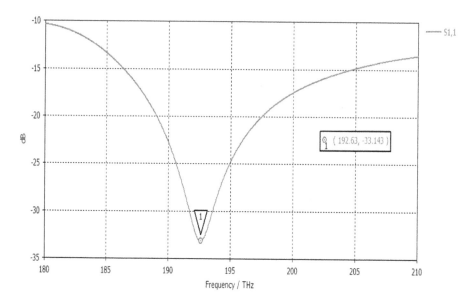

FIGURE 8.9 Reflection coefficient of nano CNDRA at 193 GHz.

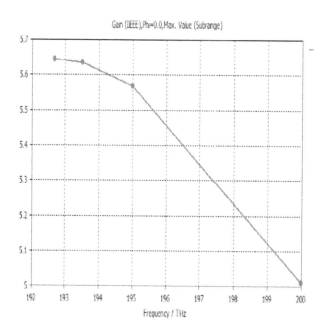

FIGURE 8.10 Gain 5.6 dBi of nano CDRA at 193 GHz.

NANO DRA   E- FIELDS

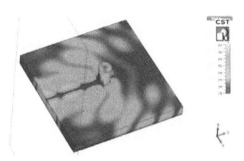

FIGURE 8.11   Currents in CNDRA.

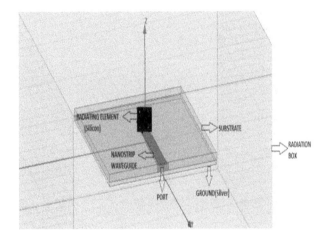

FIGURE 8.12   NDRA.

results of quantum antenna as labeled in each figure. Figures 8.12 through 8.20 are antenna results as captioned in each figure.

A collision integral can provide an exact solution of transport pentameter conditions of NDRA solution of PEC/PMC. This will result in excitation of even or odd number of modes, i.e. fundamental and higher order modes. Work is inspired from the natural phenomenon of photosynthesis. Dissipation factor or dielectric loss can be predicted based on the Drude model. Photon generation can be called fermionic or Bosonic fields. Beam formation can be worked based on correlation coefficients, and radiation or emissions are based on the Dirac second quantized field equation. Feynman path integral is used to define photon density and moments fluctuations.

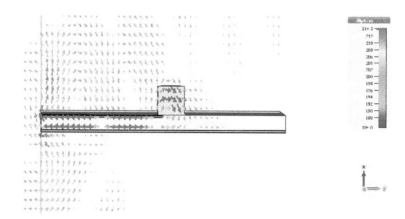

FIGURE 8.13   NDRA fields.

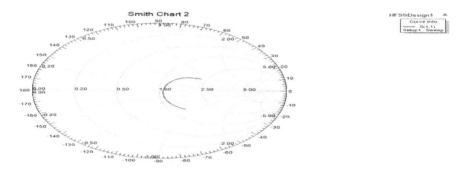

FIGURE 8.14   NDRA impedance given by Smith chart.

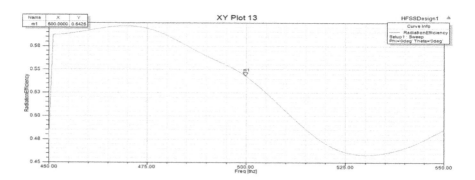

FIGURE 8.15   Radiation efficiency of NDRA.

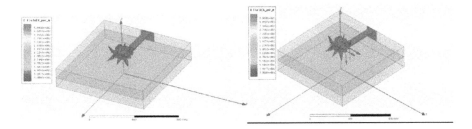

FIGURE 8.16   Electric and magnetic fields in NDRA.

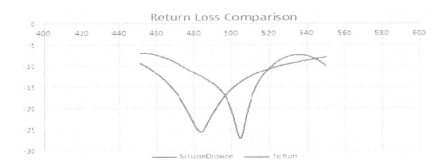

FIGURE 8.17   Reflection coefficients of NDRA at 483 and 502 THz with SiO_2 and Teflon substrates.

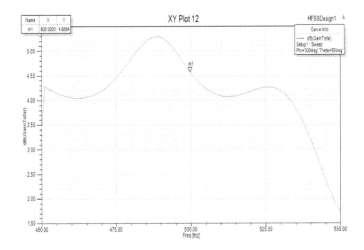

FIGURE 8.18   Gain of NDRA more than 4.5 dBi at 483 and 502 THz.

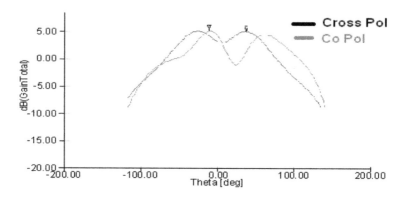

FIGURE 8.19   Radiation pattern plots of co- and cross-polarization at 483 GHz.

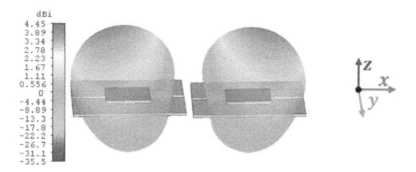

FIGURE 8.20   MIMO grapheme antenna radiation pattern at 1.82 THz at Port-1 and Port-2.

Photon spin is the main cause of nonlinearity in quantum antennas. Beaming of these nonlinear photons fields requires treatment of quantum mechanics for getting a solution of transmitting quantum states from one location to another. The Dirac equation can provide non linear solution to radiations in photonic antennas. This current density becomes the main to introduction of quantum electromagnetic field. This field can be estimated based on a solution of retarded potential. A solution of resonant modes can be worked out making use of linear super position of photon fields by creation and annihilation. Here we apply the standard commutation rule for photons and the anti-commutations rule for electron, positron/hole.

8.3   MATHEMATICAL FORMULATIONS

The following sequence of activities takes place is quantum antennas.

1. Light in light out.
2. Input excitations is Gaussian beam from LASER.
3. Nano waveguide feed is used at optical frequency.

4. SPP takes place at SiO_2 substrate and ground metal interface.
5. Nano DRA is excited by SPP through proximity feed due to capacitive coupling.
6. Radiated frequency is lower to input laser frequency.
7. Electrons, positrons. and photon fields become radiated beams of quantum antenna at terahertz frequencies with clouds of photon spin fields becoming an input excitation (SPP) to nano DRA.

This is a very important and exciting phenomenon in quantum antenna. Equations 8.1 through 8.21 describe the mathematical aspect of its working principle, i.e., input excitation to radiated power of quantum antenna.

$$\lambda_{eff} = n_1 + n_2 \left(\frac{\lambda}{\lambda_p} \right) \tag{8.1}$$

where:

λ_p is the plasma wavelength

n_1, n_2 are the geometric constants

$\lambda - \lambda_{eff}$ is the geometric factor (range 2.5) for most of metals in optical antennas

Electro-optical antenna (carbon metal used) SPR-coupling (Electron-Plasmon) design and optimization of nonlinear antennas is a challenge. The input is a single nonlinear transmission, i.e., multiple modes beam milling/lithography. In coherently coupled photon sources, direction of emission is steered using phase control mechanism. Feed gap used for impedance matching. Optical antenna operate on "light in light out" mode. The photons radiations are due to photo current introduced into optical antenna.

$$\in(\omega) = \in'(\omega) + i'' \in (\omega) \in (\omega) = 1 - \frac{\omega_p}{\omega^2 + i\Gamma\omega'}$$

$$\omega_p = \frac{\sqrt{n.e^2}}{m.\in_0} \tag{8.2}$$

$$\nabla \times \nabla \times E(r,\omega) - \frac{\omega^2}{c^2} \in (r,\omega) E(r,\omega) = 0$$

$$\rho = \in 0 E_z e^i \left(k_x^x - \omega t \right)$$

$$\text{Mean/Fluctuations} = \frac{\left\langle \psi \left| A_u(x) A_v(x') \right| \psi \right\rangle}{\left\langle \psi \left| Au1(x_1) Au2(x_2) \dots Aun(x_n) \right| \psi \right\rangle}$$

$$\text{where } A_u(x) = \frac{\mu 0}{4\pi} \int \frac{\hat{J}_u(p)e^{ipx}}{p^2 - i0} dp^4$$

$$h \cdot c / \lambda = E; \text{ Photon energy} \tag{8.3}$$

where:
   ω_P is the plasma volume frequency
   Γ is the damping constant
   ω is the center frequency

$$\omega > \omega_P$$

The Drude model is used to characterize metal transport properties when switching from microwave to optical frequency. It is dispersive in nature. Loss tangent is dielectric dissipation factor

$$\text{Tan } \delta = \frac{\epsilon''}{\epsilon'} \tag{8.4}$$

At THz frequency, penetration of electromagnetic fields into the metal ground plane takes place. This causes electron oscillations or plasmonic resonance or surface wave plasmonic resonance. Skin depth is a function of frequency, resistance, and conductance.

$$\delta = \sqrt{\frac{2\rho}{\omega\mu}} \sqrt{\sqrt{1 + (\rho\omega\epsilon)^2} + \rho\omega\epsilon} \tag{8.5}$$

$$Z_0 = \sqrt{\frac{j\omega_0 \mu_0 \mu_{vz}}{\sigma 0}}; \text{ Surface impedance} \tag{8.6}$$

$$Z = \frac{P}{|I|^2} \tag{8.7}$$

Silver nano waveguide dimensions are given as:

$$L(\lambda_0) = \frac{m \lambda_{eff}}{2}, \ m - \text{Resonant modes order} \tag{8.8}$$

$$\lambda_{eff} = \frac{2L(\lambda_0)}{m} = \frac{\lambda_0}{\eta_{eff}}$$

The length of nano waveguide will be shorter at optical frequency as compared to microwaves. The transcendental equation gives a solution to determine resonant frequency.

$$k_z \tan\left(k_z z / 2\right) = \sqrt{\left(\epsilon_r - 1\right)k_0^2 - k_z^2}\, ; \text{ when propagating in } z \text{ direction} \qquad (8.9)$$

Resonant frequency is given by transcendental equation

$$k_0^2 = \sqrt{k_x^2 + k_y^2 + k_z^2}\, ; \qquad (8.10)$$

where $k_0^2 = {}^{\omega}\sqrt{\mu_0 \,\epsilon_0}$ and $k_x = \dfrac{m\pi}{a}; k_y = \dfrac{n\pi}{b}; k_0 = 2\pi/\lambda;$

$$\sigma = \epsilon_0 \epsilon_z \, e^i \left(k_x^x - \omega t\right) \qquad (8.11)$$

Surface change density is given as below:

$$\rho(t,r) = -e\left|\psi\left(t,r\right)\right|^2 \qquad (8.12)$$

Current density is given below as:

$$\vec{J}(t,r) = -\frac{i\,e\,h}{2m}\left[\psi\left(t,r\right)\nabla\vec{\psi}\left(t,r\right) - \vec{\psi}\left(t,r\right)\nabla\psi\left(t,r\right)\right]; \qquad (8.13)$$

$\psi\left(t,r\right) \rightarrow$ Wave function

$$i\,[\,\frac{\partial\psi\left(t,r\right)}{dt} = \frac{l^2}{2m}\nabla^2\psi\left(t,r\right) - Z\frac{e^2}{r}\psi\left(t,r\right)$$

Electric scalar vector potential is given as:

$$\varnothing(t,r) = -\frac{e}{4\pi\varepsilon 0}\int\frac{\psi\left(t - \dfrac{\left|r - r'\right|}{C},\ r'\right)}{\left|r - r'\right|}d^3 r' \qquad (8.14)$$

Magnetic vector potential is:

$$A(t,r) = -\frac{i\,e h\mu_0}{8\pi\,m}\int\psi\left(\pm\frac{r - r'}{C}, r\right)X\nabla\vec{\psi}\left|t - \frac{r - r'}{C}r'\right|\nabla'_\psi\left(t\frac{r - r'}{C}\right)d_r^3 \qquad (8.15)$$

$\psi(t,r)$ satisfies Schrodinger's equation, which gives a solution of wave function for single electrons:

$$\frac{\partial \psi(t,r)}{dt} = \frac{h^2}{2m}\nabla^2_\psi(t,r) - \frac{Ze^2}{r}\psi(t,r) \tag{8.16}$$

This provides the discrete dynamic solution.

If there are N electrons in the joint wave function, then the wave function can be described as:

$$\frac{\partial \psi}{\partial t} = -\frac{h^2}{im}\sum_{1=1}^{N}\nabla^2\psi - \sum_{1=1}^{N}\frac{Ze^2}{\left|n\right|}\psi \tag{8.17}$$

Hence, current density can be expressed as:

$$J(x) = e\psi(x)^* \lambda^\mu \psi(s) \tag{8.18}$$

where $\lambda^\mu = \gamma^0\gamma^\mu$.

This satisfies the Dirac second quantized field equations and is obtained in the quadratic function of the Dirac field operators. The current density produces a quantum electromagnetic field.

Quantum retarded potential:

$$A(t,r) = \int \frac{J^\mu\left(t - \left|r - r'\right|, r'\right)}{4\pi\left|r - r'\right|}d^3r' \tag{8.19}$$

where $A(t,r)$ is the Bosonic field. It is expressed as quadratic function field operators.

Quantum antenna radiated fields:

$$\left\langle \psi \left| A_u(x) A_v(x') \right| \psi \right\rangle \quad \text{Fluctuation fields} \tag{8.20}$$

$$A_u(x) = \frac{\mu_0}{4\pi}\int \frac{\hat{J}_\mu(p)e^{ipx}}{p^2 - 10}d_P^4 \tag{8.21}$$

Equations 8.1 through 8.21 have described the mathematical part of NDRA right from resonant frequency to currents and power radiated by the THz antenna.

8.4 RESULT AND DESCRIPTION OF NDRA

Different models at sub-wavelengths have been developed and simulated on different geometries to get different resonant frequencies in the THz range. These have been realized as NDRA, and Figures 8.12 through 8.21 show the results of NDRA under

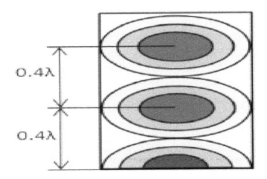

FIGURE 8.21   Magnetic dipole formation inside NDRA.

excitation for getting radiation results. CST software has been used for getting radiation parameters as shown in Figures 8.1 and 8.2. Hardware of these models could not be developed because of nonavailability of fabrication and measuring facilities at present. These simulated results have opened vistas of opportunities to the new field of quantum antenna for 5G and 6G mobile communication and biomedical and high-resolution scanning applications.

With the illumination of femtosecond laser pulses, electron-hole pairs are created in the photoconductive material when the laser pulses have higher photon energy than the band gap energy of the photoconductive material. The quick change in the electric field at the gap results in a transient current, and finally an electromagnetic pulse in the THz frequency range is radiated. Mean of photons along with spin field included during radiation need to develop quantum mechanics control for desired radiation orientation characteristics. A silver nanostrip or nano waveguide is used in the model. The model with design dimensions is given below:

8.4.1 Nano DRA Applications

There are many advanced applications of quantum antennas at THz frequencies such as:

1. Imaging for pattern recognition with higher resolution of images, such as biosensor spectroscopy.
2. Large band width for 5G mobile communication.
3. Compact size for wearable, secured communication due to encryption.
4. Higher order modes for dimension scaling, i.e., miniaturization.
5. Defense applications, i.e., detection of size of aircraft.
6. On chip communication explosive detection.
7. Detection of any metallic object carried by any person at airport without manual search
8. At airport, advanced landing system during bad weather.
9. Rectenna for energy harvesting, i.e., renewable energy source. Nano antenna can receive direct optical energy.

10. Biological sensor detection.
11. Comprehensive integrated border management system.
12. Rectennas work the whole day (day and night) and are not sensitive to the weather conditions. Applications of rectennas can be Drones, e-cars, and sensors.

8.5   CONCLUSION

NDRAs have a very sharp beam width and high gain. They are compact and have a very high operating frequency, i.e., THz frequency. They are a good candidate for high-resolution imaging and scanning. New fabrication and testing facilities at such a high frequency in the THz range are needed. Correlation coefficient manipulation is used in beam formation in quantum antenna. It is observed that spin of photons becomes the main function of nonlinearity at optical frequency. There is a strong need to establish a relationship between radiated frequencies, surface plasmonic positrons (SPP) resonance frequencies, and input excitation frequency. The SPP wavelength is ten times smaller to free space wavelength, so the radiated frequency shall be smaller to input excitation frequency in plasmonic antennas.

Terahertz antenna can be very useful in medical sensing, ultra fast data transfer and wide band applications. The source of coherent pulsed THz radiation is photoconductive antennas. The ultra fast variations of surface photo conductivity of a semiconductor substrate near optical excitation by LASER introduces surface plasmon resonance (SPR). After exciting the gap between the electrodes by LASER, the concentration of charge carriers increases into the graphene substrate, being THG (third harmonic generation), and THz pulse generation occurs. Graphene is an inhomogeneous material. Multimodal NLO imaging is an excellent tool in biomedical applications. Using nonlinear optics and Maxwell's equations, a set of nonlinear wave equations is derived to give scattered electromagnetic fields from an inhomogeneous medium. Multiple nonlinear interaction in the medium also gives multiple modes and treating multimodes with convolution gives wide bandwidth. GaAs material introduces to second harmonic generation (SHG) material, and graphene with SiO_2 substrate introduces to third harmonic generation (THG) material. The laser excitation is used to exploit several nonlinear optical effects that can provide high-contrast imaging of biological samples. Third harmonic generation (THG) can create light with three times the energy of the incident light. THG exploits a nonlinear interaction between high intensity light from a laser and a material. Graphene has strong light matter interaction and a strong *third* order nonlinear response.

Terahertz waves are absolutely harmless to humans and have no ionizing radiation, unlike X-ray machines. A sequence of microwave pulses is injected into a particular point on an object. The transmitted or reflected portion of each pulse is then detected after a time delay. Terahertz light is dominated by the dielectric response of the material's different absorption features compared with healthy tissue. The THz body scanner advantages include terahertz imaging cameras and THz sources. *Nonlinear optical* (NLO) imaging is a powerful microscopy technique, i.e., multimodal NLO imaging is an excellent tool in biomedical applications. The electromagnetic imaging may be perceived as the retrieval of the spatial distribution of the permittivity function of the medium. Here, measurement of the scattered fields interacts with the medium. The frequency of the illuminating electromagnetic wave may vary from GHz to THz range. A nonlinear Riccati-type differential equation would be derived, which represents the fundamental equation of one-dimensional imaging. Resolution degradation due to bandwidth limitations is resolved by using antenna arrays. Terahertz radiation can pass through clothing, paper, cardboard, wood, masonry, plastic and ceramics. The penetration depth is typically less than that of microwave radiation. Terahertz radiation has limited penetration through fog and clouds and cannot penetrate liquid water or metal. Currently medical imaging depends on terahertz imaging, but terahertz imaging is still a very immature field. By varying the length of this delay, a 3D image of the internal structure can be discerned. The object may be translated through the beam or the beam may be scanned across the object to build up an image at various points.

We have identified dental imaging. Currently dentists try to identify if a patient has tooth decay using visual inspection or X-ray radiography. THz imaging has high-resolution due to which, early stage detection of tooth decay will become a possible solution. Terahertz imaging can also measure the absorption spectrum at each pixel, which could then be used to assess the effects of a reagent on the water content of the skin. The THz regime of the electromagnetic spectrum is rich with the emerging possibilities in imaging applications with unique characteristics for screening for weapons, explosives and biohazards and imaging of concealed objects, water content and skin, and these advantages can be harnessed by using the effective THz sources and detectors.

In THz imaging systems, the pulsed THz sources and detectors find unique applications and thus we have emphasized revisiting these kinds of systems.

Several novel imaging techniques which exploit the distinctive properties of the THz systems have been presented. Moreover, the THz antenna is one of the most important components of a THz imaging system as it plays a significant role in both impedance matching and power source.

Therefore, the recent developments in THz antenna design for imaging applications are reviewed and the potential challenges of such THz systems are investigated. The photoconductive antennas form the basis of many THz imaging and spectroscopy systems and find promising applications in various scientific fields. However, for the imaging applications, there is a requirement of planar and compact THz antenna sources with on-chip fabrication and high directivity in order to achieve large depth-of-field for better image resolution. Therefore, the key modalities of improving photoconductive dipole antenna performance are identified for imaging applications. Also, the ways to improve the directivity of the photoconductive dipole antenna are discussed. THz antenna technology has effective solution for imaging and scanning applications.

Short pulse lasers generate wideband systems, reliable for through-clothes imaging of person-borne concealed objects. Using the frequency-modulated continuous-wave (FMCW) radar technique with a nearly 30 GHz bandwidth, subcentimeter range resolution is achieved. To optimize the radar's range resolution, a reliable software calibration procedure compensates for signal distortion from radar

waveform nonlinearities. The clutter (clothing over skin) behaves like a rough surface scatterer and can be characterized by a normalized backscatter terahertz-based
stand-off security screening system ("body scanner"). The concept of a body scanner
meant for detection of objects concealed under clothes is briefly described on our
web page dedicated to security for homes. The Terasense body scanner is intended
for stand-off detection of weapons, including cold steel and firearms, bombs and gre-
nades, explosive belts, and various contraband items hidden under clothes. The effective imaging range (imaging distance) of up to 3 m ensures remote detection of
suspicious objects hidden under clothes. The mere presence of such hidden items
may help law enforcement and security personnel to mark a suspect person and iso-
late him/her for subsequent physical inspection (body search) by police/security/
customs officers—depending on the scene.

- Panchromatic-absorption image: This is constructed by measuring the
  absorption coefficient, $\alpha(\omega)$, over the entire frequency bandwidth of the
  terahertz pulse. $\alpha(\omega)$ is determined by measuring the change in the amplitude of either the transmitted or reflected terahertz radiation.
- Monochromatic-absorption image: The absorption coefficient can also be
  measured at a fixed frequency or over a limited frequency range covered by
  the terahertz pulse.
- Time-of-flight image: The thickness of the object can be determined by
  measuring the time delay, D, for the radiation to travel through the object
  compared with free space. The thickness, t, is then calculated according to
  the simple relation $t = (n-1)/Dc$, where n is the refractive index of the portion of the object sampled by the pulse and c is the speed of light.
- Refractive-index image: The average refractive index, at either a fixed
  frequency or over a range of frequencies, can be calculated from a portion of an object with uniform thickness using the simple relation above.
  The application of these antennas opens up new vistas of opportunities for
  providing alternate resources to E-vehicles, large bandwidth for 5G mobile
  applications, compact size and high-resolution imaging and scanning for
  biomedical applications. Quantum antenna can provide DC voltages by
  receiving e.m. waves from satellites.

REFERENCES

1. R. S. Yaduvanshi and H. Parthasarthy, *Rectangular DRA Theory and Design*. New Delhi, India: Springer, 2016.
2. M. Hayasi, *Quantum Information Theory*. Berlin, Germany: Springer, 2017.
3. C. A. Balanis, *Antenna Theory: Analysis and Design*. New York: Wiley, 2005.
4. A. Öchsner and H. Altenbach, *Properties and Characterization of Modern Materials (Advanced Structured Materials)*. Singapore: Springer, 2016.
5. S. A. Long, *Dielectric Resonator Antenna Handbook*. Boston, MA: Artech House, 2007.

6. P. Bhardwaj, B. Deutsch and L. Novotny, "Optical Antennas," *Adv. Opt. Photon.*, vol. 1, pp. 438–483, 2009.

7. D. H. Youn, M. Seol, J. Y. Kim, J. W. Jang, Y. Choi, K. Yong and J. S. Lee, "TiN Nanoparticles on CNT–Graphene Hybrid Support as Noble-Metal-Free Counter Electrode for Quantum-Dot-Sensitized Solar Cells," *Chem Sus Chem*, vol. 6, no. 2, pp. 261–267, 2013.

8. R. P. Feynman and A. R. Hibbs *Quantum Mechanics and Path Integration.* 1965.

9. L. Novotny and N. Van Hulst, "Antennas for Light," *Nat. Photonics*, vol. 5, no. 2, pp. 83, 2011.

10. W. T. Sethi, H. Vettikalladi and H. Fathallah, "Dielectric Resonator Nano Antenna at Optical Frequencies," In *Information and Communication Technology Research (ICTRC), 2015 International Conference on*, pp. 132–135. IEEE.

11. F. J. Rodríguez-Fortuño, A. Espinosa-Soria, and A. Martínez, "Exploiting Metamaterials, Plasmonics and Nanoantennas Concepts in Silicon Photonics," *J. Opt.*, vol. 18, no. 12, pp. 123001.

12. L. Zou, W. Withayachumnankul, C. M. Shah, A. Mitchell, M. Bhaskaran, S. Sriram and C. Fumeaux, "Dielectric Resonator Nano Antennas at Visible Frequencies," *Opt. Express*, vol. 21, no. 1, pp. 1344–1352, 2013.

13. X. Wu, F. Tian, W. Wang, J. Chen, M. Wu and J. X. Zhao, "Fabrication of Highly Fluorescent Graphene Quantum Dots Using L-Glutamic Acid for In Vitro/ In Vivo Imaging and Sensing," *J. Mater. Chem. C*, vol. 1, no. 31, pp. 4676–4684, 2013.

14. A. E. Krasnok, D. S. Filonov, C. R. Simovski, Y. S. Kivshar and P. A. Belov, "Experimental Demonstration of Superdirective Dielectric Antenna," *Appl. Phys. Lett.*, vol. 104, no. 13, pp. 133502.

15. N. Shinohara and H. Matsumoto "Experimental Study of Large Rectenna Array for Microwave Energy Transmission," *IEEE Trans. Microw. Theory Tech.*, vol. 46, no. 3, pp. 261–268.

16. G. Varshney, A. Verma, V. S. Pandey, R. S. Yaduvanshi and R. Bala, "A Proximity Coupled Wideband Graphene Antenna with the Generation of Higher Order TM Modes for THz Applications," *Opt. Mater.*, vol. 85, pp. 456–463, 2018.

17. S. Abadal, E. Alarcón, A. Cabellos-Aparicio, M. Lemme and M. Nemirovsky, "Graphene-Enabled Wireless Communication for Massive Multicore Architectures," *IEEE Commun. Mag.*, vol. 51, no. 11, pp. 137–143, 2013.

18. (a) M. Schnell, P. Alonso-González, L. Arzubiaga, F. Casanova, L. E. Hueso, A. Chuvilin and R. Hillenbrand, "Nanofocusing of Mid-Infrared Energy with Tapered Transmission Lines," *Nat. Photonics*, vol. 5, no. 5, pp. 283–287, 2011. (b) J. Wen, S. Romanov and U. Peschel, "Excitation of Plasmonic Gap Waveguides by Nanoantennas," *Opt. Express*, vol. 17, no. 8, pp. 5925, 2009.

19. J.-S. Huang, T. Feichtner, P. Biagioni and B. Hecht, "Impedance Matching and Emission Properties of Optical Antennas in a Nanophotonic Circuit," *Nano Lett.*, vol. 9, no. 5, pp. 1897–1902, 2008.

20. L. Zou, W. Withayachumnankul, C. M. Shah, A. Mitchell, M. Klemm, M. Bhaskaran and S. Sriram, "Efficiency and Scalability of Dielectric Resonator Antennas at Optical Frequencies," *IEEE Photonics J.*, vol. 6, no. 4, pp. 1–10, 2014.

21. W. Chen, R. L. Nelson, D. C. Abeysinghe and Q. Zhan, "Optimal Plasmon Focusing with Spatial Polarization Engineering," *Opt. Photonics News*, pp. 36–41, 2009.

22. P. Biagioni, J.-S. Huang and B. Hecht, "Adult Teratoma of the Testicle Metastasizing as Adult Teratoma," *Reports Prog. Phys.*, vol. 75, pp. 1–40, 2012.
23. Y. Zhao and A. Alu, "Optical Nanoantennas and Their Applications," *IEEE Radio Wirel. Symp. RWS*, pp. 58–60, 2013.
24. G. N. Malheiros-Silveira, G. S. Wiederhecker and H. E. Hernández-Figueroa, "Dielectric Resonator Antenna for Applications in Nanophotonics," *Opt. Express*, vol. 21, no. 1, pp. 1234–1239, 2013.

9 MIMO DRA

9.1 INTRODUCTION

Multi-input multi-output dielectric resonator antennas (MIMO DRAs) work based on the analogy of throwing a stone into a pond and creating traveling water waves, doing so repeatedly to create more waves, thus generating a pattern of waves into the pond. It means modulation of amplitude of water waves takes place. The amplitude of water waves gets increased or decreased as time during propagation of water waves. It is also a function of the origin of wave to stones thrown into the pond. MIMO antennas for communication systems work like the above stated analogy. MIMO antennas have few very important parameters such as envelopment correlation coefficient (ECC), directive gain (DG), mean effective gain (MEG), channel capacity loss (CCL) and total active reflection coefficient (TARC) [1–5]. In this chapter, the proof of concept on various MIMO antenna has been developed for practical applications. DRA has been arranged with feed network to work as MIMO. Two port and four port MIMO has been built and a theoretical mathematical concept has been developed. In this chapter, development of microwave, millimeter waves and terahertz frequency antennas is discussed with prototypes. Simulation and experimental results of different MIMO DRAs has been investigated and discussed.

If the phenomenon of throwing multiple stones into a pond at same location when hit from the same direction can create coherence and synchronism, then the magnitude of water waves will be enhanced as weighted sum response of water waves [5–10]. This is how the MIMO antenna operates in any communications system to introduce for large gain. Similarly, if all stones are thrown at different locations in the pond, there will be generation of many waves and each wave can carry different information independently, but for less distance due to low magnitude. Large data can be transported in that case. MIMO diversity performance parameters like ECC, diversity gain (DG), mean effective gain (MEG), channel capacity loss (CCL), and total active reflection coefficient (TARC) with their simulated results have been obtained.

In this concept combination above, two statements can be made to achieve large data rates and long distance. That is why MIMO and massive MIMO are the most popular in today's antenna technology [11–15]. The modern communication system requires high data transmission rates, which can be achieved using systems with high bandwidth or multi-input and multi-output (MIMO) systems [1].

A number of research papers based on the microstrip patch MIMO antennas have been reported in the literature [2]. The limitation of the microstrip patch-based antennas is that the performance of these antennas is weak at higher frequencies due to conductor and surface wave losses [3]. The dielectric resonator (DR) antenna (DRA) provides wide bandwidth, high gain and significant radiation efficiency. These prominent features of the DRA encouraged researchers to use it in MIMO systems. Some of the DR-based MIMO antennas have been reported in the literature [4–8]. Figures 9.1 through 9.26 have presented different MIMO designs and their results obtained with simulated/measured parameters.

The necessary points about the MIMO DRA like the multiple radiators must have a common reference plane [9]. In this paper, a MIMO DRA is presented with half-conical shaped DR elements. The MIMO antenna structures are designed and numerically analyzed. The antenna shows high isolation between each port and a low envelop correlation coefficient, which confirms acceptable MIMO performance [9]. These MIMO DRAs can be utilized for single, dual band and triple operation in microwave and higher frequency ranges of terahertz [16–21].

Envelope correlation coefficient "ρ" shows the influence of different propagation paths of the RF signals that reach the antenna elements. An ECC equation can also be given based on radiation pattern.

Here, Figure 9.1 through 9.3 shows the arrangement of MIMO antenna structures.

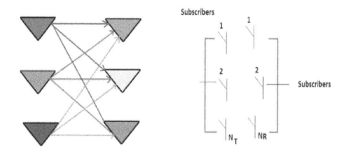

FIGURE 9.1 MIMO trans and receive concept.

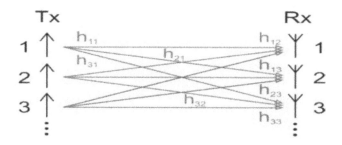

FIGURE 9.2 MIMO antenna structure.

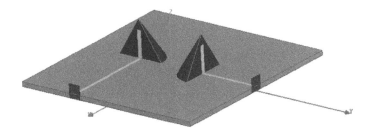

FIGURE 9.3 Half-conical MIMO DRA.

The lower the correlation, the higher data throughput. ECC, diversity gain, bandwidth, isolation and polarization are a few important parameters of MIMO antenna.

9.2 MIMO DRA WORKING: CASE-1

A wideband, half-conical shaped dielectric resonator multi-input multi-output antenna with two element configuration is proposed in this paper. Two radiators are arranged orthogonally to improve the isolation between ports. Antenna provides the dual band response with the impedance bandwidth of 28.13% (8.40–11.16 GHz) and 34.48% (7.85–11.15 GHz) in lower band through port-1 and port-2, respectively. The antenna provides the impedance bandwidth of 21.55% (11.85–14.65 GHz) and 25.33% (11.55–14.9 GHz) in the upper band through port-1 and port-2, respectively. Moreover, the low envelope correlation coefficient (ECC < 0.3) is achieved in the pass bands of the antenna.

This DRA can be visualized as a cone which has been cut vertically and split it into two equal parts with radius ($R = 10$ mm) and height ($H = 10$ mm). The cone is made up of ceramic material with dielectric constant ($\epsilon_r = 10.2$). The substrate of the antenna is of dimension $80 \times 80 \times 1.67$ mm^3 having material FR–epoxy with permittivity $\epsilon_s = 4.4$. The proposed antenna design is simulated and analyzed using a high frequency structure simulator (HFSS). Two radiators are arranged orthogonally to improve the isolation between ports. Antenna provides the dual band response with the impedance bandwidth of 28.13% (8.40–11.16 GHz) and 34.48% (7.85–11.15 GHz) in lower band through port-1 and port-2, respectively. The antenna provides an impedance bandwidth of 21.55% (11.85–14.65 GHz) and 25.33% (11.55–14.9 GHz) in the upper band through port-1 and port-2, respectively.

The half-cones are equally placed about the center of the plane. The antenna is excited by a conformal strip which is attached to the flat faces of half cones and is connected to a 50Ω microstrip line. The dimensions of feed are given as width ($w_f = 1.6$ mm) and length ($L_{f1} = 40$ mm and $L_{f2} = 25$ mm). Height of the feed at the half split cone surface is ($h_f = 10$ mm). The inter element spacing ($s = 10$ mm) decides the isolation of proposed MIMO DRA.

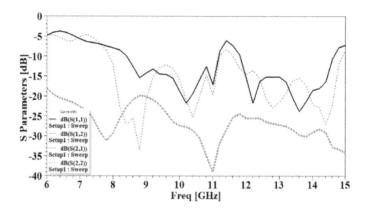

FIGURE 9.4 S_{11} reflection coefficient simulated.

9.2.1 RESULTS AND DISCUSSION

Figure 9.4 shows the S_{11} parameter response of the proposed antenna. The antenna provides a dual band response with an impedance bandwidth of 28.13% (8.40–11.16 GHz) and 34.48% (7.85–11.15 GHz) in lower band through port-1 and port-2, respectively. The antenna provides an impedance bandwidth of 21.55% (11.85–14.65 GHz) and 25.33% (11.55–14.9 GHz) in the upper band through port-1 and port-2, respectively. It can be observed that the isolation between ports of the antenna remains more than 20 dB in both of the operating pass bands. Figure 9.5 shows the calculated ECC plot of the antenna. The ECC remains less than 0.003 in both pass bands of the antenna. The requirement of the antenna of having high isolation and low ECC is fulfilled, so the proposed antenna can be a good candidate for MIMO applications (Figures 9.5).

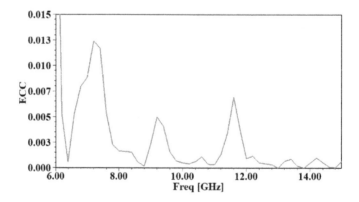

FIGURE 9.5 Simulated ECC plot of the proposed MIMO antenna.

9.3   MIMO GRAPHENE WORKING: CASE-1

The graphene two port MIMO antenna for THz communication has been discussed with low correlation coefficient and high isolation. Graphene provides miniaturization of the MIMO antenna without affecting its isolation. Mutual coupling
is reduced by placing the antenna systems on top and bottom. The compactness
of the antenna has been achieved using both sides of the substrate. The antenna
covers a frequency range of up to 0.55 THz with port 1 and 0.45 THz with port 2.
The isolation obtained is greater than 12 dB, and the error correlation coefficient remains below 0.002. It provides a diversity gain around 10 dB (Figures 9.6
through 9.9, Table 9.1).

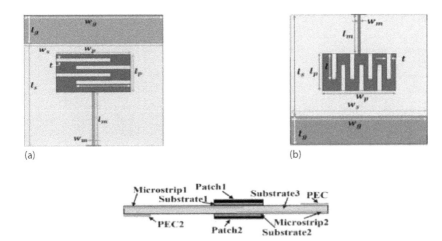

FIGURE 9.6   Meander line MIMO graphene antenna design. Layout of the proposed MIMO
antenna (a) top side and (b) bottom side.

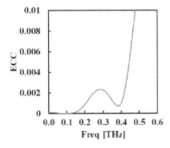 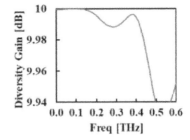

FIGURE 9.7   ECC and DG plots.

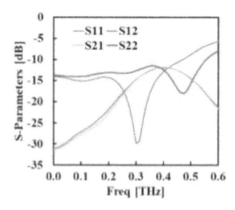

FIGURE 9.8   S-parameters simulated.

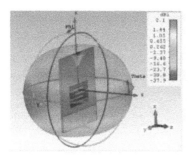
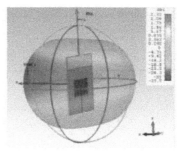

FIGURE 9.9   Radiation pattern simulated.

TABLE 9.1

Meander Line MIMO Graphene Antenna Design Dimensions

Dimensions of the Proposed Graphene Based MIMO System

Length of the ground (l_g)	30 µm
Width of the ground (w_g)	120 µm
Length of the base substrate (L_s)	140 µm
Width of the base substrate (w_s)	120 µm
Length of upper substrate (l_{s1})	40 µm
Width of upper substrate (w_{s1})	64 µm
Length of the patch (l_p)	40 µm
Width of the patch (w_p)	64 µm
Thickness of the slots (t)	4 µm
Length of the horizontal slots (l_{hs})	47 µm
Length of the vertical slots (l_{vs})	28 µm

9.4 TWO PORT THz MIMO DRA: CASE-1

The proximity coupled graphene patch two-port multi-input multi-output (MIMO) antenna is developed for THz applications. The envelop correlation coefficient of the antenna remains less than 0.01. The utilization of graphene material provides flexibility in tuning the antenna response along with conformal designs for biomedical applications. Isolation of more than 25 dB is achieved between the ports of the antenna (Figure 9.10 and Table 9.2).

ECC and conductivity of MIMO (Figure 9.11):

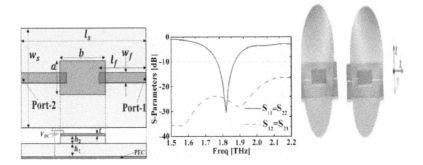

FIGURE 9.10 Graphene MIMO antenna at 1.82 THz (S-parameters and radiation pattern).

TABLE 9.2
Design Dimensions

The Antenna Dimensions (Unit of dimensions μm, except t in nm)

Parameter	l_s	w_s	h_1	h_2	l_f	w_f	a	b	t
Dimension	60	40	16	16	23	18	34	16	1

$$\rho_{eij} = \frac{|S_{11}^* S_{12} + S_{21}^* S_{22}|^2}{(1 - (|S_{11}|^2 + |S_{21}|^2))((1 - (|S_{22}|^2 + |S_{12}|^2))}$$

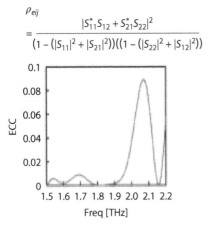

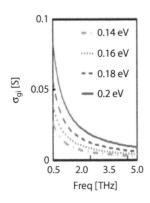

FIGURE 9.11 DRA MIMO two port results.

9.5   TWO PORT P SHAPE MIMO DRA: CASE-1

The P shape MIMO DRA utilizes the combination of half-cylinders and a slab of long rectangular DRA integrated. This shape of DRA has generated orthogonal modes with circular polarization. It provides polarization diversity in two-port MIMO DRA. The MIMO and diversity performance are investigated by calculating the performance parameters such as envelop correlation coefficient, diversity gain, mean effective gain, channel capacity loss, and total active reflection coefficient (Figures 9.12 through 9.15).

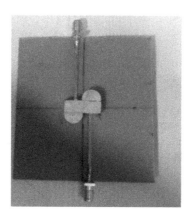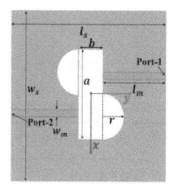

FIGURE 9.12   Two port MIMO DRA antenna. Design dimensions are: $l_s = 80$, $w_s = 80$, $l_m = 37$, $w_m = 1.6$, $a = 20.2$, $b = 6$, $r = 5.2$, $d = 5$ and $h_p = 6.6$ (the unit of the dimensions is *mm*).

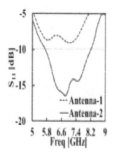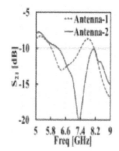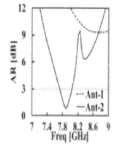

FIGURE 9.13   Resonant modes excited into MIMO DRA.

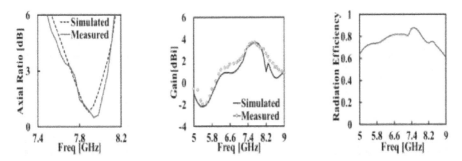

FIGURE 9.14 Gain, AR and efficiency of MIMO DRA.

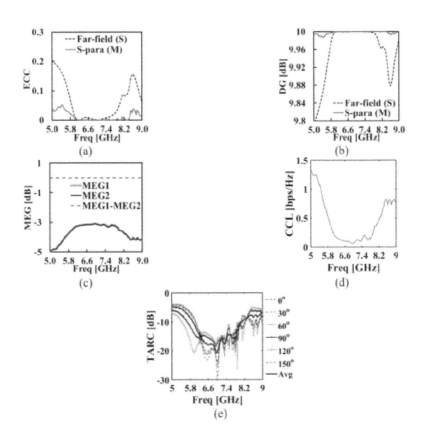

FIGURE 9.15 (a) ECC, (b) DG, (c) MEG, (d) CCL, and (e) TARC.

9.6 FOUR PORT MIMO DRA: CASE 1

The MIMO DRA has a structure of two epsilon-shaped DRAs integrated with two cylindrical DRAs. The hybrid of modes $\mathbf{TM_{01\delta}}$ and $\mathbf{HEM_{31\delta}}$ is generated. Creation of slots in the ground plane has fetched isolation between ports. The radiation pattern of the antenna can be steered in different directions with the excitation at different ports. The antenna provides 15.73% overlapping 10 dB impedance bandwidth with t gain 4–6 and 6–6.5 dB at port-1/2 and port-3/4, respectively. The antenna has efficiency of more than 93% at port-1, 2 and 97% at port-3/4. The envelop correlation coefficient and diversity gain are within the acceptable limits (Figure 9.16 through 9.19). The design dimensions are given below:

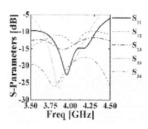
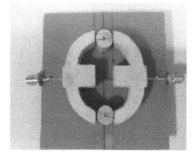
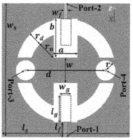

FIGURE 9.16 Four port Epsilon shaped MIMO antenna. $l_s = 80$, $w_s = 80$, $l_f = 15$, $w_f = 1.6$, $a = 12$, $b = 19$, $r = 5$, $w = 12$, $d =$, $l_g = 15$, $w_g = 7$, $r_a = 11$, $r_d = 7.3$, $h = 10$, $h_f = 8$ (the unit of the dimensions is mm).

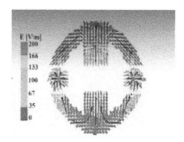

FIGURE 9.17 Four port DRA MIMO results.

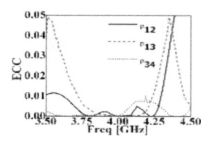
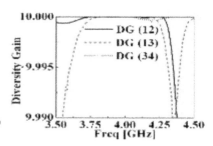

FIGURE 9.18 Four port DRA MIMO results.

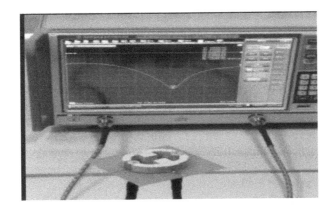

FIGURE 9.19   MIMO four port VNA measurements.

9.7   TWO PORT MIMO DRA: CASE-1

Figure 9.20 is MIMO structure and Figure 9.21 is S_{11}.

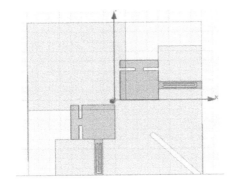

FIGURE 9.20   Two port MIMO antenna.

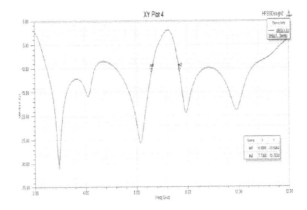

FIGURE 9.21 S_{11} antenna (2–12 GHz).

9.8 MIMO DRA: CASE-1

The DRA MIMO is tuned to obtain the triple/dual/single band operation. A triple band antenna is designed with two radiating elements. Slots of size equivalent to quarter wavelength are introduced in the substrate and ground plane of the antenna. Introduction of slots converts the triple band response of the antenna into wide dual band response. Furthermore, changing the dimensions and relative permittivity of the radiating elements leads to merge both bands, and a wide single band response is obtained. The isolation between ports can also be improved using the antenna configuration with orthogonal feeding. The MIMO and diversity performance parameters like error correlation coefficient, diversity gain, channel capacity loss, mean effective gain and total active reflection coefficient have also been investigated (Figures 9.22 through 9.26).

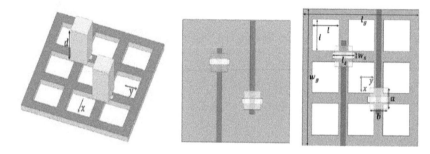

FIGURE 9.22   MIMO DRA.

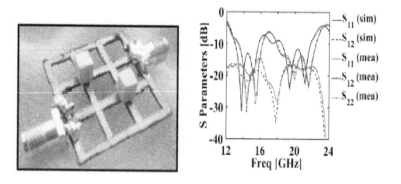

FIGURE 9.23   Hardware design of MIMO DRA and S_{11}.

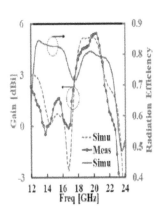
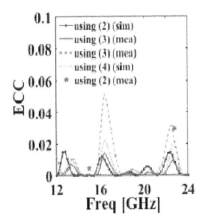

FIGURE 9.24   Gain of MIMO and ECC.

FIGURE 9.25   MIMO I DRA two port inside anechoic chamber for radiation pattern measurements.

FIGURE 9.26   MIMO DRA four port inside anechoic chamber for radiation pattern measurements.

9.9　CONCLUSION

Two port and four port MIMO DRAs have been analyzed with their working models. Prototype models have been tested for s-parameters with VNA and anechoic chamber for desired performance check. MIMO DRAs have been developed for microwave, millimetric wave and terahertz applications. DRA MIMO has been tested in anechoic chambers for stable results. Graphene antenna are conformal and very compact in size. Hence, these antenna can be used for biomedical and space applications. Near future sapphire MIMO can be developed for aesthetic and robust designs. Vehicular antennas can also be designed with a combination of meander lines and DRAs for various applications. Multi port input and multi port output can increase diversity into communication systems.

REFERENCES

1. A. Goldsmith, *Wireless Communications*. Cambridge, MA: Cambridge University Press, 2005.
2. M. A. Jensen and J. W. Wallace, "A Review of Antennas and Propagation for MIMO Wireless Communications," *IEEE Trans. Antennas Propagat.*, vol. 52, no. 11, pp. 2810–2824, 2004.
3. D. M. Pozar, *Microwave Engineering*, 4th ed. Hoboken, NJ: John Wiley & Sons, 2005.
4. A. Sharma and A. Biswas, "Wideband Multiple-Input–Multiple-Output Dielectric Resonator Antenna," *IET Microwaves, Antennas Propag.*, vol. 11, no. 4, pp. 496–502, 2017.
5. A. A. Khan, R. Khan, S. Aqeel and J. Nasir, "Dual-Port MIMO DRA with High Isolation for WiMAX Application," *3rd International Conference on Engineering & Emerging Technologies (ICEET)*, 2016.
6. A. A. Khan, R. Khan, S. Aqeel, J. U. R. Kazim, J. Saleem and M. K. Owais, "Dual-Band MIMO Rectangular Dielectric Resonator Antenna with High Port Isolation For LTE Applications," *Microw. Opt. Technol. Lett.*, vol. 59, no. 1, pp. 44–49, 2016.
7. A. A. Khan, R. Khan, S. Aqeel, J. Nasir, J. Saleem and Owais, "Design of a Dual-Band MIMO Dielectric Resonator Antenna with High Port Isolation for WiMAX and WLAN Applications," *Prog. Electromagn. Res. M*, vol. 50, pp. 65–73, 2016.
8. M. S. Sharawi, S. K. Podilchak, M. U. Khan, and Y. M. Antar, "Dual-Frequency DRA-Based MIMO Antenna System for Wireless Access Points," *IET Microwaves, Antennas Propag.*, vol. 11, no. 8, pp. 1174–1182, 2017.
9. M. S. Sharawi, "Current Misuses and Future Prospects for Printed Multiple-Input, Multiple-Output Antenna Systems," *IEEE Antenna Propag. Mag.*, 2017.
10. J. D. Gale et al., "The Rise of Graphene," *Rev. Mod. Phys.*, vol. 58, no. 1, pp. 710–734, 2012.
11. A. H. Castro Neto, F. Guinea, N. M. R. Peres, K. S. Novoselov, and A. K. Geim, "The Electronic Properties of Graphene," *Rev. Mod. Phys.*, vol. 81, no. 1, pp. 109–162, 2009.
12. K. S. Novoselov et al., "Electric Field Effect in Atomically Thin Carbon Films," *Science*, vol. 306, no. 5696, pp. 666–669, 2004.
13. D. Tse, D. Tse, P. Viswanath, and P. Viswanath, "Book—Viswanath & Tse," *Notes*, p. 583, 2004.
14. A. Akdagli and A. Toktas, "Compact Multiple-Input Multiple-Output Antenna with Low Correlation for Ultra-Wide-Band Applications," *IET Microwaves, Antennas Propag.*, vol. 9, pp. 822–829, 2015.

15. A. A. Ibrahim, M. A. Abdalla, A. B. Abdel-Rahman, and H. F. A. Hamed, "Compact MIMO Antenna with Optimized Mutual Coupling Reduction Using DGS," *Int. J. Microw. Wirel. Technol.*, vol. 6, no. 2, pp. 173–180, 2014.

16. J. M. Jornet and I. F. Akyildiz, "Graphene-Based Nano-Antennas for Electromagnetic Nanocommunications in the Terahertz Band," *Proc. Fourth Eur. Conf. Antennas Propag.*, pp. 1–5, 2010.

17. Z. Xu, X. Dong, and J. Bornemann, "Design of a Reconfigurable MIMO System for THz Communications Based on Graphene Antennas," *IEEE Trans. Terahertz Sci. Technol.*, vol. 4, no. 5, pp. 609–617, 2014.

18. M. S. Khan, A. D. Capobianco, S. M. Asif, A. Iftikhar, B. D. Braaten, and R. M. Shubair, "A Properties Comparison Between Copper and Graphene-Based UWB MIMO Planar Antennas," *2016 IEEE Antennas Propag. Soc. Int. Symp. APSURSI 2016 - Proc.*, pp. 1767–1768, 2016.

19. I. F. Akyildiz and J. M. Jornet, "Realizing Ultra-Massive MIMO (1024Ã—1024) Communication in the (0.06–10) Terahertz Band," *Nano Commun. Netw.*, vol. 8, pp. 46–54, 2016.

20. M. S. Sharawi, "Printed MIMO Antenna Systems: Performance Metrics, Implementations and Challenges," *Forum for Electromagnetic Research Methods and Application Technologies (FERMAT)*, no. 1, pp. 1–10, 2014.

21. C. M. Luo, J. S. Hong, and L. L. Zhong, "Isolation Enhancement of a Very Compact UWB-MIMO Slot Antenna with Two Defected Ground Structures," *IEEE Antennas Wirel. Propag. Lett.*, vol. 14, no. c, pp. 1766–1769, 2015.

10 Horn DRA Antenna

10.1 INTRODUCTION

The dielectric resonator material used in resonant antennas was proposed in 1983. Due to minimized metallic loss, the dielectric resonator antenna (DRA) is efficient particularly as compared to metallic antennas if operated at millimeter wave and terahertz frequency spectrums. High dielectric constant material can further reduce the size of the DRA. Hence, it is a small and low profile antenna. Low cost dielectric materials are now easily available commercially. The horn antenna is a type of aperture antenna. The radiation fields from aperture antenna can be determined from fields over the aperture. The aperture fields become the sources of the radiated fields at far fields, and the frequency spectrum is classified as LF, HF, VHF, UHF, SHF, L, S, C, X, Ku, K, Ka, mm wave, THz, etc. There are different types of DRAs, such as E-plane, H-plane, and EH or pyramidal horn. The horn antenna is designed by flaring a hollow rectangular cross section to a larger opening. Horn antennas are easy to excite for providing high gain and have a wide impedance-bandwidth, implying that the input impedance is fairly constant over a wide frequency range. A long horn with small flare angle is required to obtain uniform aperture fields distribution. Both fundamental and higher order modes have been obtained. Mode merging can provide large bandwidth, and mode shifting can result in multi bands. Higher order modes can result in higher gain [1–11].

10.2 HORN ANTENNA MAIN PARAMETERS

The horn antenna is generally used for testing in these frequencies for the parameters of antenna given as follows:

1. Frequency
2. Impedance
3. Power
4. Gain
5. Reflection coefficient (s_{11})
6. VSWR
7. Bandwidth
8. Beam width
9. Directivity
10. Polarization
11. Axial ratio bandwidth
12. Impedance bandwidth
13. Frequency ratio

14. Directive gain and power gain
15. Correlation coefficient
16. Circular polarization
17. RHCP and LHCP

The frequency spectrum of horn DRA is given in Table 10.1.
    An important parameter for horn DRA is its resonant design frequency:

$$f_r = \frac{c}{2\pi\sqrt{\varepsilon_r}}\sqrt{2\left(\frac{\pi}{a}\right)^2 + \left(\frac{\pi}{2b}\right)^2}$$

where:
f_r is the resonant frequency of TE_{nmp} – mode,
a,b is the dimensions of the DRA,
ε_r is the permittivity of the dielectric material used in DRA, and
c is the velocity of light ($\approx 3\times10^8$ m/s).

Quality factor of Horn DRA:

$$Q(\omega) = \frac{2\omega \cdot W(\omega)}{|I(\omega)|^2 R_r(\omega)}$$

where:
ω is the resonant frequency,
R_r is the radiation resistance, and
W is the energy stored.

TABLE 10.1

Microwave Frequency Bands

Letter Designation	Frequency Range
L band	1–2 GHz
S band	2–4 GHz
C band	4–8 GHz
X band	8–12 GHz
K_u band	12–18 GHz
K band	18–26.5 GHz
K_a band	26.5–40 GHz
Q band	30–50 GHz
U band	40–60 GHz
V band	50–75 GHz
E band	60–90 GHz
W band	75–110 GHz
F band	90–140 GHz
D band	110–170 GHz

The quality factor Q of rectangular DRA can also be evaluated by comparing the power radiated $P_{rad} = \frac{1}{2}I^2 R_r$, and the average electromagnetic energy (W) stored with the horn DRA is given as follows.

$$W(\omega) = \frac{1}{4}\int_{(0,a)(0,b)(o,d)}\left[\epsilon\left(E,E'\right) + \mu\left(H,H'\right)\right]d_x d_y d_z$$

Average energy stored per unit cycle with the horn DRA is given by,

$$P(\omega) = \frac{\omega W(\omega)}{2\pi} = \frac{\omega W(\omega)}{2\pi}$$

Short magnetic dipoles are formed in any antenna. They can be classified as electromagnetic devices designed for radar and any communication system. Antenna has basic theory for EM wave radiations (based on second order differential equations). Generation of EM wave in DRA takes place due to acceleration and deceleration of charge careers. Device for energy transformation into space and vice versa (electrical signal to radio wave and vice versa).

- Radiation
- Electrical energy
- Magnetic energy
- Dipole formation
- Dipole moments
- Wave propagation based on Helmholtz and Maxwell's equations
- Frequency is determined by the characteristic equation and more accurate by the transcendental equation
- Concept of electrical length due to fringing fields is applied for accurate calculations of resonant frequency
- Transcendental equation solution for taking electrical length account
- PEC, PMC walls concept (boundary value problems)
- Modes of propagation due to perturbation into standing waves
- High value of dielectric constants of different values is available due to different DR materials
- $\lambda = \frac{C}{F}$; Dimensions of DRA are $= \frac{\lambda g}{\sqrt{\epsilon_r}}$
- Acceleration or second order differential equation solution for wave function (time-space relationship)

Dielectric horn DRA with aperture is coupled with linearly polarized (LP) radiation patterns horizontally and vertically. Horn antennas are widely used in areas of wireless communications, sensing, testing of radio frequency communication, and radar systems. They are used as high gain elements in phased arrays antennas. Horn antenna also act as feed elements for satellite and radar reflectors and serve as a universal standard antenna for calibration and gain measurements. Gain is closer to directivity of DRA horn antenna because of high efficiency. Polarized radiation patterns have low cross-coupling, and a multifunction circular polarized dielectric horn with the z slot or staircase-shaped slot is realized for left-hand circular polarized (LHCP) and right-hand circular polarized (RHCP). In the third design, along with the circular polarization, the gain is enhanced using the parasitic patches on the superstrate. Pyramidal horns are antennas of high gain and can operate over a wide range of frequencies; therefore, they are used in numerous applications. The pyramidal horn antennas have flaring opened in both E- and H-plane directions and are suitable for high gain, high power and broadband applications due to low level back radiation and good directivity and gain characteristics. Dr. Jagdish Chandra Bose coined the term horn "collecting funnel" and developed 60 Ghz horn antenna in 1987. Pyramidal horn antennas are used because they have low crossover polarization losses and increased radiation efficiency due to their simplicity and versatility and large gain and relative of large gain. In this chapter, we have integrated the rectangular dielectric resonator antenna with the pyramidal horn. The DRA operated at 2.4 GHz is used to excite the horn antenna. DRA offers various advantages such as small size, light weight, flexible excitation scheme and relatively wide bandwidth when compared with patch antennas as low metallic and surface wave losses. The dielectric resonator is simply made of dielectric that are low. The aperture coupled DRA is chosen because its feed network is isolated from the DRA, thus spurious radiation is avoided. When the DRA is excited through slot, it acts like a horizontal magnetic dipole and minimizes cross polarized fields. The DRA is operated at TE113 higher order mode. As we know, DRA is affected by its aspect ratios when the aspects ratios are so chosen that other modes close to the operating mode are excited. Generally, higher order modes increase bandwidth and gain of the antenna. In this proposed design, the DRA is fed to the horn for high gain and wideband operation. A horn DRA diagram is shown in Figure 10.1.

For a given gain (G) and operating frequency (f) with a and b dimensions of feed or wave guide the design procedure is shown in Figure 10.2.

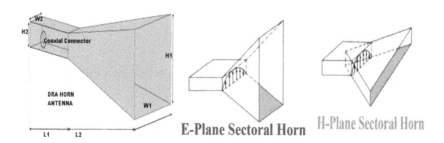

FIGURE 10.1   Horn DRA with E- and H-plane.

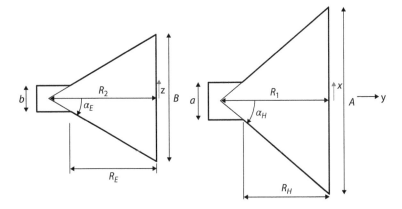

FIGURE 10.2 E and H part of horn antenna.

Thus, we have system of four equations with four unknowns (A, B, R_1, R_2) (Table 10.2 and Figures 10.3 through 10.8).

$$R_1 \mid R_H = A \mid (A - a);$$

$$R_2 \mid R_E = B \mid (B - b);$$

The wave guide part of the horn antenna is rectangular and the opening of the horn is pyramid-shaped, which makes it convenient for impedance match to free space. Due to the horn section, the beam shape can be achieved due to its focal point properties. The horn acts as a concave lens to the beam.

TABLE 10.2
Survey Work on Horn Antenna

Sl. No	Year	Author	Horn	Results
1.	2005	Nasimuddin K. P. Esselle	A rectangular dielectric resonator integrated with quasi-planar surface mounted horn.	The antenna gain of 8.5 dB at 6.0 GHz, giving a 4.9 dB increase compared to the standard dielectric-resonator antenna (DRA).
2.	2007	Nasimuddin K. P. Esselle	Dielectric resonators and surface mounted short horns (SMSH).	Gain is enhanced by 3.7 to 9.8 dB at 5.95 GHz, bore sight cross-polarization level is less than −27 dB.

(Continued)

TABLE 10.2 *(Continued)*
Survey Work on Horn Antenna

Sl. No	Year	Author	Horn	Results
3.	2007	Nasimuddin Karu Esselle	Wide band hybrid DRA integrated on patch (DRoP) with a surface-mounted short horn (SMSH).	Gain greater than 9 dBi within the 2:1 VSWR bandwidth 28% at 6.04 to 8 Ghz due to DRA on patch (DRoP).
4.	2010	A. Othman, M. F. Ain, A. A. Sulaiman, M. A. Othman	Ka-band horn antenna excited with parasitic dielectric resonator antenna.	Excitation DRA produced $HEM_{11\delta}$ mode, high gain of 19.8 dB.
5.	2010	Mohd Fadzil Ain, Zainal Arifin Ahmad, Mohamadariff Othman, Ihsan Ahmad Zubir	Parasitic cylindrical dielectric resonator excited horn antenna at 38 GHz. The horn antenna is fed by five identical cylindrical parasitic dielectric resonators.	The DRA excited $HEM_{11\delta}$ mode, high gain of 19.8 dB.
6.	2013	Ranjana Manohar Makam	Pyramidal horn antenna excited with various top-hat loaded monopoles.	The directivity is 15.3 dB.
7.	2015	Chintan A. Patel Shobhit K. Patel	Pyramidal horn antenna design loaded by meta-material split ring resonator SRR.	Due to SRR, the gain is increased to 14 dB.
8.	2017	Ravi Dutt Gupta Manoj Singh Parihar	Differentially fed wideband rectangular DRA with high gain horn.	Impedance bandwidth of 30% and gain of 12.2 dB.

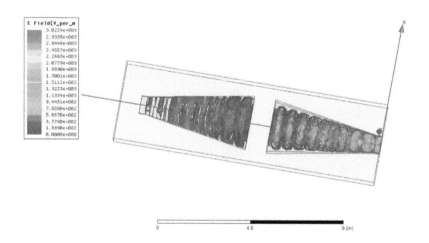

FIGURE 10.3 E-field pattern of horn antenna.

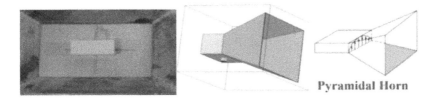

FIGURE 10.4   Pyramidal-shaped DRA horn antenna.

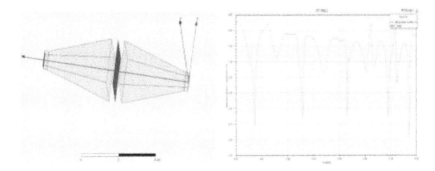

FIGURE 10.5 S_{11} measurement results, when one metal strip is placed in between two horn DRAs.

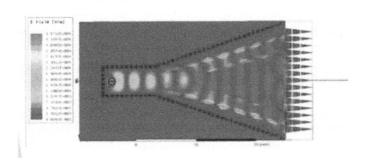

FIGURE 10.6   E-field pattern of SIW horn antenna for millimeter wave frequency.

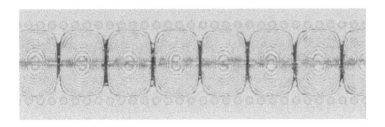

FIGURE 10.7   H-field pattern of SIW horn antenna for millimeter wave frequency.

S11 Reflection Coefficient

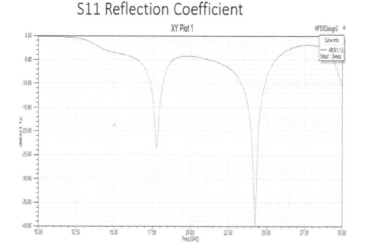

FIGURE 10.8 S_{11} of SIW horn antenna for millimeter wave frequency.

Quality factor is an important parameter of horn DRA.

Gain is closer to directivity of horn antenna because of high efficiency.

10.3   HORN DRA DESIGNS

10.3.1   Slot Aperture Horn DRA Design Issues

The slot aperture coupling can be conveniently integrated on a substrate. The slot length should be large enough to achieve sufficient coupling to DRA through feed. The large aperture DRA resonate in the designed frequency band. The slot is resonant in nature. Hence, its size is frequency dependent for excellent electromagnetic fields coupling to the DRA. As a rule of thumb, good starting values of slot length l_s and slot width w_s are $L_s = 0.4\,\lambda_0/\varepsilon_{e\,ws} = 0.2 \times L_s$, i.e., λ_0 is the free-space wavelength, $\varepsilon_{eff} = (\varepsilon_r + \varepsilon_s)/2$, ε_r and ε_s are the relative dielectric permittivity of the DRA and substrate, respectively.

    In this chapter, a 2.4 GHz pyramidal horn has been designed, and total experimental and mathematical analysis of the same is performed. This is designed as a slot fed DRA in horn shape. The horn was designed using rectangular dielectric waveguide, which has dimensions 95.25×42.33 mm^2 and thickness of 2.032 mm. The horn antenna acts as a resonator without any additional resonator. Hence, it results in wideband applications. The horn antenna has HE_{mn} hybrid modes and a combination of the both TE and TM modes inside it. At the aperture of the horn, the wave takes the spherical wave front, and hence it has higher gain and directivity with narrow beam width. The DRA is designed using material of 30 permittivity. It operates at 2.4 Ghz. The DRA operating mode is TE_{113} mode. The rectangular slot-fed DRA is also excited at a higher-order mode. The advantage of slot fed DRA is that it avoids spurious radiations. The DRA is placed over the ground plane. Thus by using the image theory, the height of DRA becomes twice (2d). The slot is centrally located beneath the DRA and it is perfectly matched to 50 ohm impedance. In this condition, only odd modes are excited into horn DRA. The aspect ratio of the DRA is responsible for modes excitation. The fundamental TE_{111} mode and the higher-order TE_{113} modes have broadside radiation patterns. The higher mode has various advantages such as higher gain and bandwidth. As we know in DRA, for even and odd modes solution the tangential components of electric fields and normal components of magnetic fields must satisfy required boundary conditions. These are symmetric to the middle plane, which acts as a magnetic wall. TE mode is an odd number ($m = 1,3,5...$). In case of odd mode, the index m tangential component electric field is asymmetric about the middle plane, which is the electric wall. Thus, DRA boundary conditions decide which modes to excite. From these we can conclude that for TE mnp, both $x = 0$, and $z = 0$ planes are electric walls, when m, n are odd numbers. The $y = 0$ plane is a magnetic wall when p is an odd number. Modes in this design cannot be excited if p is an even number. For exciting an even number of modes, DRA should be isolated, i.e., without ground plane in this case.

10.3.2 PYRAMIDAL HORN

The pyramidal horn was designed using the waveguide as feed and its operating frequency at 2.4 GHz (Figure 10.9).

The parameters of horn DRA will depend upon the following dimensions:

Aperture width, aperture height, waveguide width, waveguide height, waveguide length, horn length, probe length, probe radius, and distance of feed (Tables 10.3 and 10.4).

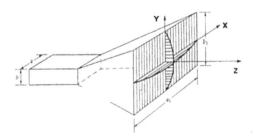

FIGURE 10.9   Pyramidal horn DRA.

TABLE 10.3
Design Dimension of Pyramid Horn DRA

Design	Dimension (in mm)
WAVEGUIDE "a" DIMENSION in xy plane	95.25 mm
WAVEGUIDE "b" DIMENSION in xy plane	42.33 mm
WAVEGUIDE LENGTH in xz plane	105.83 mm
HORN SIZE "A" in xy plane	190.5 mm
HORN SIZE "B" in xy plane	148.17 mm
HORN FLARE LENGTH in yz and xz plane	317.5 mm
WALL THICKNESS	2.032 mm

TABLE 10.4
Design Dimensions of Trapezoidal DRA Horn Antenna

	Design Parameters	Dimensions
1.	INSIDE TRAPEZIODAL DRA (Taconic TMM = 10)	LOWER RECTANGLE LT1 × WT1 = 50 mm × 25 mm UPPER RECTANGLE LT2 × WT2 = 100 mm × 50 mm HEIGHT = 42.25 mm
2.	OUTSIDE TRAPEZIODAL DRA (Glass = 5.5)	LOWER RECTANGLE LG1 × WG1 = 62 mm × 35 mm UPPER RECTANGLE LG2 × WG2 = 110 mm × 60 mm HEIGHT = 42.25 mm
3.	SUBSTRATE	Lsub × Wsub × Hsub = 200 mm × 200 mm × 0.8 mm
4.	MICROSTRIP LINE	Lm × Wm = 100 mm × 2.46 mm
5.	STUB	Lstub × Wstub = 6.0232 mm × 2.46 mm
6.	SLOT	LS × WS = 20.24 mm × 4.04 mm
7.	GROUND PLANE	Lg × Wg = 200 mm × 200 mm

10.3.3 Design Example Using HFSS (Figures 10.10 through 10.15)

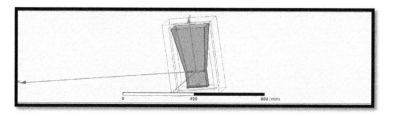

FIGURE 10.10 Design of pyramidal horn integrated of rectangular dielectric resonator feed at 2.4 GHz.

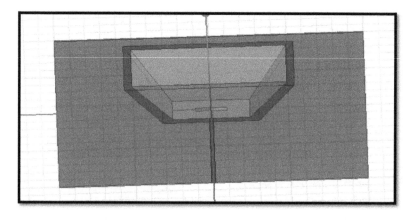

FIGURE 10.11 Trapezoidal horn DRA design with dimensions as per Table 10.3.

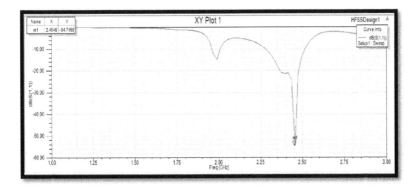

FIGURE 10.12 S_{11} plot of horn DRA.

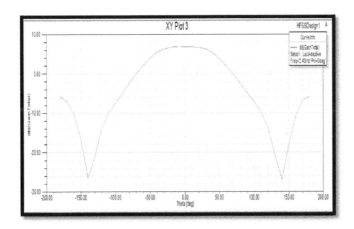

FIGURE 10.13 Gain plot of DRA.

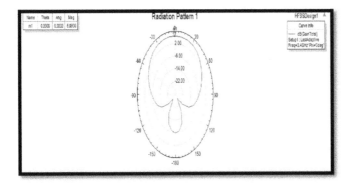

FIGURE 10.14 Radiation pattern of horn DRA.

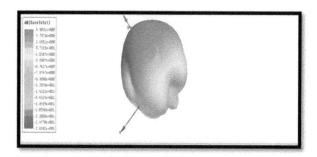

FIGURE 10.15   Radiation pattern 3D plot of horn DRA.

10.3.4 TRAPEZOIDAL HORN DIELECTRIC RESONATOR ANTENNA DESIGN

The rectangular DRA of dielectric material, $E_r = 30$ is used. The microstrip feed line is designed on the FR4 epoxy having dielectric constant, $E_r = 4.4$, a thickness of substrate, $d = 0.8$. The design dimensions are $120 \times 70 \times 2.032$ mm^3. Simulation results of S_{11}, Z_{11}, gain, radiation pattern, E-fields, H-fields and VSWR has been obtained and presented. Design dimensions are placed in Table 10.4. Circular polarization is also achieved in the horn. Superstrate trapezoidal horn DRA has the most advantage in gain enhancement (Figure 10.16). Figures 10.17 through 10.31 have obtained design and simulation results of trapezoidal horn DRA as stated in proper labels (Figure 10.32 and Table 10.5).

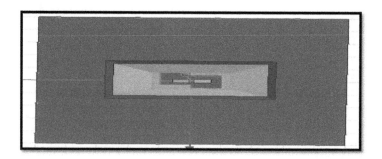

FIGURE 10.16   Trapezoidal horn DRA with Z-shaped insert.

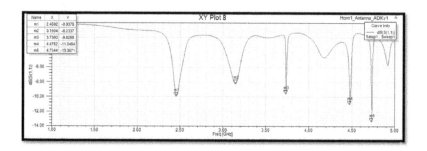

FIGURE 10.17 S_{11} simulated reflection coefficient of trapezoidal horn DRA.

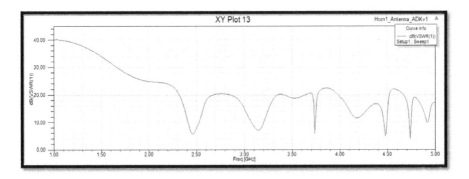

FIGURE 10.18   Simulated VSWR of trapezoidal horn DRA.

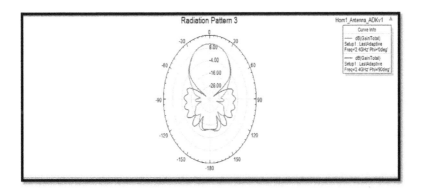

FIGURE 10.19   Simulated radiation pattern gain plot of trapezoidal horn DRA.

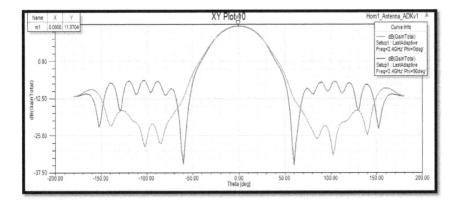

FIGURE 10.20   Simulated 2D gain plot vs theta at phi = 0 shows the magnetic field and 90 showing the electric field gain is 11 dbi.

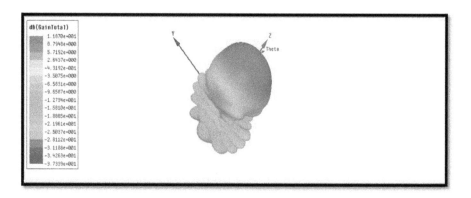

FIGURE 10.21 Simulated 3D polar plot.

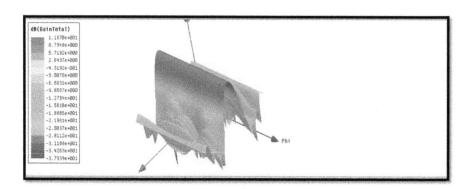

FIGURE 10.22 Simulated 2D gain plot.

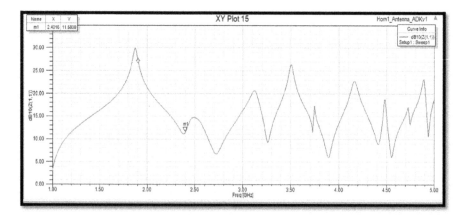

FIGURE 10.23 Simulated Z_{11} vs frequency plot.

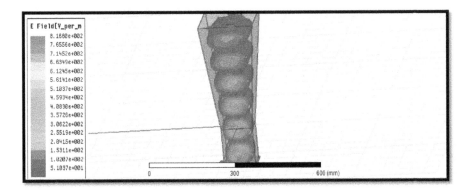

FIGURE 10.24   Simulated H-fields inside the horn showing higher order modes.

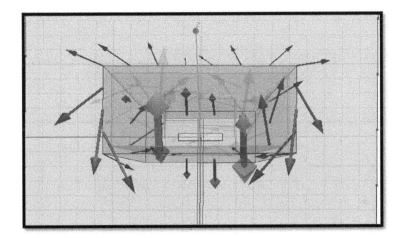

FIGURE 10.25   E-field vector of horn DRA.

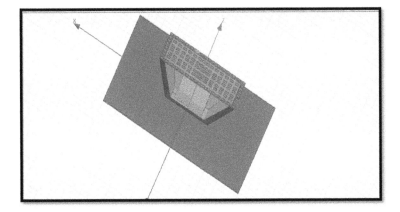

FIGURE 10.26   Horn DRA with superstrate.

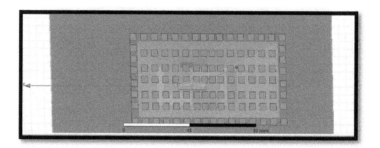

FIGURE 10.27   Superstrate placed on top of horn DRA.

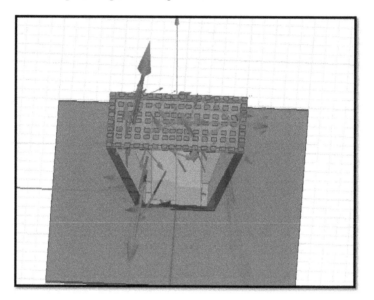

FIGURE 10.28   Horn DRA E-fields.

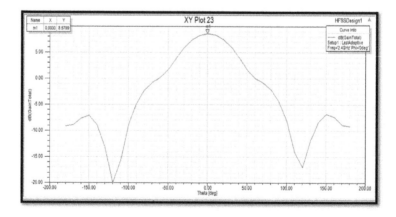

FIGURE 10.29   Horn DRA 2D gain plot.

E-Fields

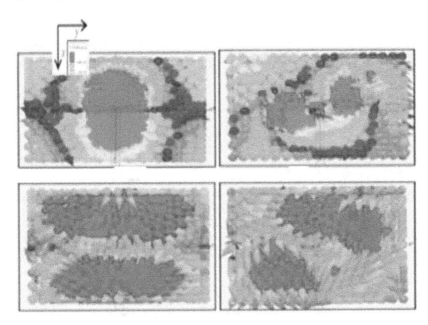

FIGURE 10.30 Circular polarized fields due z-shaped material insert into horn DRA.

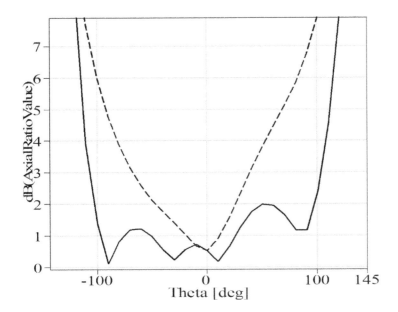

FIGURE 10.31 Axil ratio of trapezoidal horn DRA at 2.4 Ghz.

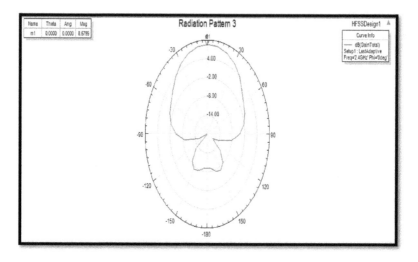

FIGURE 10.32 Trapezoidal horn DRA radiation pattern plot at 2.4 Ghz.

TABLE 10.5
Trapezoidal DRA Horn Design

Design Parameter	Dimension (in mm)
DRA length in xy plane	22
DRA width in xy plane	12.5
DRA height in xz plane	17
Microstrip line length	41.3
Microstrip line width	2.6
Substrate height in xy plane	0.8
Substrate length in xz plane	120
Substrate width in xy plane	70
Slot length in xy plane	11
Slot width in xy plane	2.6

10.4 HORN DRA WITH RECTANGULAR SUPERSTRATE

Far field and near field description:

Typical antenna test range measurement accuracies are shown in Tables 10.6 and 10.7.

Beam prediction: $\leq 1\%$ of 3 dB beam width

Minimum distance between transmit and receiver for far filed measurement:

$$D\,los = 2D^2/\lambda$$

TABLE 10.6
Specifications

Gain:	±0.2 dB, (on relative measurement)
	±0.5 dB, (on absolute measurement)
Side lobe:	±1.5 dB, at 35 dB below peak
	±3.0 dB, at 50 dB, below peak
Cross Pol.	±1.5 dB, at 35 dB below peak

TABLE 10.7
Technical Specifications

Freq. (GHz)	λ (m)	D (m)	$2D^2/\lambda$ (m)
1	0.3	20	2666.67
1	0.3	3	60
10	0.03	2	266.667
40	0.005	1	400

where:
  D is the largest dimension of Ant and
  λ is the wavelength.

1. Reactive near-field $= r < \lambda 2\pi$.
2. Electromagnetic energy in this region varies very rapidly with distance.
3. Radiating near-field $= \frac{\lambda}{2\pi} < r < 2D^2/\lambda$.
4. Energy density in this region remains nearly constant with respect to the distance from the antenna. It is this region in which near field measurements are done.
5. Radiating far field $= r > 2 << 2D^2/\lambda$.
6. The energy density of electromagnetic waves decays very rapidly according to the inverse square law as a function of distance.
   D is the size of the antenna aperture.
   r is the distance from the antenna aperture.

10.5   CONCLUSION

Horn DRA is a highly-matched, highly directive, versatile application with stable output and high gain. It is used as a universal antenna in tests and evaluation of prototypes and other RF devices to gain measurements as well as feed horn to reflector based antenna. The wide examples are parabolic bellbird case grain antennas, satellite antennas and high gain radar antennas. Multiple modes in horn DRA thus can be excited, and high gain is obtained. The design and performance of dielectric resonator integrated with pyramidal horn have been presented. The simulation results obtained are showing wide bandwidth and excellent gain. The radiation characteristics of E and H have been presented. The gain of antenna is considerably increased to 11.8 dBi, and the axial ratio is below 3 dB in the horn DRA. It offers circular polarization, and the antenna can be easily adapted for high gain and various array applications. Horn DRA are considered advantageous at millimeter wavelengths due high efficiency. Multiple higher order modes in horn DRA have been obtained. At higher order modes, high antenna gain is also obtained. At higher modes, higher gain is also obtained. Further results can also be obtained in horn DRA with concave and convex superstrates for getting better directivity and control. Normally, wideband antennas have low gain, and high gain antennas have narrow bandwidth. The designed antenna operates in S-band (2.4 GHz). The DRA horns are linearly and circularly polarized. The pyramidal DRA horn is excited by feed, and the slot fed DRA is excited at higher order mode TE_{113}. This has resulted in a broadside radiation pattern. The far field radiation pattern is also obtained. E-field and H-field analysis of S-band pyramidal DRA horn antenna has been obtained with simulation.

Directivity is highly desired in vehicular and radar systems. DRA horn structure is simple and can be used in testing of RF devices.

REFERENCES

1. A. Petosa, *Dielectric Resonator Antenna Handbook*. Boston, MA: Artech House Publishers, 2007.
2. Nasimuddin, K. P. Esselle and A. K. Verma, "GAIN Enhancement of Aperture-Coupled Dielectric-Resonator Antenna with Surface Mounted Horn," in *Proc. 28th Gen. Assembly URSI*, New Delhi, India, October 23–29, 2005.
3. K. T. Selvan, "Accurate Design Method for Optimum Gain Pyramidal Horn," *Electron. Lett.*, vol. 35, pp. 249–250, 1999.
4. M. A. Othman, A. R. Othman and N. Saysoo, "Development of Ultrawide Band (UWB) Horn Antenna Using Approximation Method," *IEEE Symposium on Wireless Technology & Applications (ISWTA)*, September 23–26, 2012, Bandung, Indonesia.
5. D. M. Pozar, "Microstrip Antenna Aperture-Coupled to a Microstrip Line," *Electron. Lett.*, vol. 21, pp. 49–50, 1985.
6. Y. Pan and K. W. Leung, "Wideband Circularly Polarized Trapezoidal Dielectric Resonator Antenna," *IEEE Antennas Wirel. Propag. Lett.*, vol. 9, pp. 588–591, 2010.
7. Y. Pan, K. W. Leung and E. H. Lim, "Compact Wideband Circularly Polarised Rectangular Dielectric Resonator Antenna with Dual Underlaid Hybrid Couplers," *Microw. Opt. Technol. Lett.*, vol. 52, no. 12, pp. 2789–2791, 2010.
8. R. S. Yaduvanshi and H. Partsarthy, *Rectangular Dielectric Resonator Antenna*. New Delhi, India: Springer.

9. A. Petosa, A. Ittipiboon, Y. Antar, D. Roscoe and M. Cuhaci, "Recent Advances in Dielectric-Resonator Antenna Technology," *IEEE Antennas Propag. Mag.*, vol. 40, no. 3, pp. 35–48, 1998.

10. K. M. Luk and K. W. Leung, *Dielectric Resonator Antennas*. Baldock, UK: Research Studies Press, 2003.

11. G. Almpanis, C. Fumeaux and R. Vahldieck, "The Trapezoidal Dielectric Resonator Antenna," *IEEE Trans. Antennas Propag.*, vol. 56, no. 9, pp. 2810–2816, 2008.

11 Anechoic Chamber with Absorbers

11.1 INTRODUCTION

Anechoic chambers or shielded rooms for antenna and EMI/EMC measurements were developed and built in 1940 by Murray Hill at Bell Lakes, USA. These are rooms designed to absorb all unwanted reflections produced by a source during measurement. Thus they are echo free and shielded from the external environment for ideal characteristic measurement of any antenna parameters used for any wireless system. Electronic/electrical system using electromagnetic waves operate at low and high frequency. The size of the chamber depends on the device under test (DUT) operating characteristics, such as frequency and size. The term *reflection* less is used as figure of merit of all anechoic chambers. RF absorbers are used to absorb reflected energy. Absorbers make up 60% of the cost of any chamber. RAM (radar absorbent material) is used as an absorber. It operates on the principle resonance or resistivity. It is designed based on Faraday's cage principle. Measure of reflectivity inside any anechoic chamber is taken standard. The anechoic chamber is a large room and may be rectangular, tapered, ellipsoid or spherical. It is covered with metallic boundaries for EM shielding. These metallic boundaries are covered by RF absorbers of pyramid shapes. Absorber qualification mainly depends on operating RF power and frequency. These foam absorbers are secured with absorbing tiles placed before RF absorbers. Quiet zone is the main volume inside an anechoic chamber, which defines its efficiency compared to total size of the chamber. RF absorbers help to reduce reflections. Murray Hill, Richard Johnson, W. H. Emersion, D. Fasold, Vince Rodriguez, Koestler, Holloway, Hess, Gilreath, Dong, Hiroki, Zhang, Zhou, Farahbaksh, Evan, Togar, Ali, Dong, Ota, Oriki, Sang, Kuester, Anzai, Xie, Deng, Nornikman, Malek, Ahmed, Wee, Azremi, Hess, Yaghjian, Sun, Imran, Rizal, Lee, Burnside and Hansen are well known scientists who have contributed much to this field research since its inception [1–18]. Orbiter, MVG and Franconia are major manufacturers of these anechoic chambers. Figures 11.1 through 11.5 are about measurements of antenna parameters inside anechoic chamber.

FIGURE 11.1 Anechoic chamber measuring system.

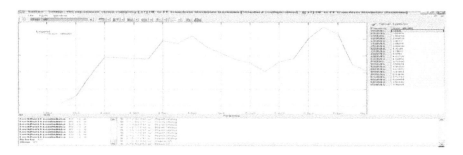

FIGURE 11.2 Measured gain in anechoic chamber.

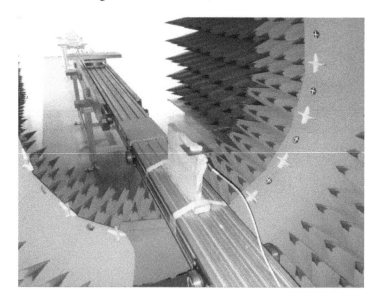

FIGURE 11.3 DRA device under test inside anechoic chamber with feed and measuring probes.

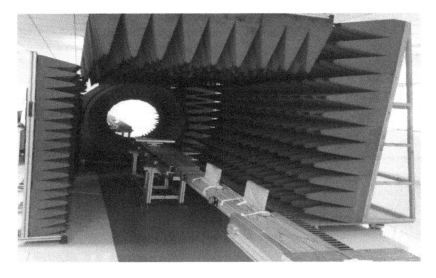

FIGURE 11.4 DRA device under test inside anechoic chamber with absorber.

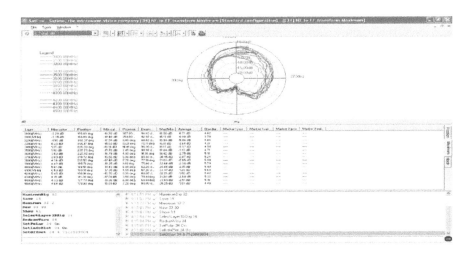

FIGURE 11.5 DRA device measured radiation pattern inside anechoic chamber.

11.2 DESCRIPTION OF ANECHOIC CHAMBER

Key subsystems of anechoic chambers are RF source, antenna, feed system, positioner or turn table, RF receiver, control automated system, Midas software, work station and plotter and device under test (test antenna). Vince Rodriguez has developed basic design guidelines of indoor anechoic chambers for antenna parameters testing. However, outdoor facilities can also be used to test these antenna parameters. These measurements are subject to external environmental conditions, i.e., external interference due to temperature, rain, humidity, etc. To avoid all these limitations, there was a strong need of indoor measurement test facilities known as shielded chambers or anechoic chambers. Measurements from these chambers were found to be error free, i.e., without any reflection and interference or distortions. Thus, people started checking anechoic chamber results as qualification of antennas, electronic and electrical systems tested are indoors, with test facility as a prerequisite in the development phase. This made anechoic chambers the most popular test facilities. These chambers are frequency, size of DUT and power dependent. Antenna parameters as qualification include frequency, impedance, power, gain, reflection coefficient (s_{11}), VSWR, bandwidth, beam width, directivity, polarization, permittivity, permeability, dialectic constant, efficiency, quality factor, resonant modes axial ratio, aspect ratio, low profile, directive gain, dipole moment, higher order modes, Poynting vector, magnetic vector potential, wave number, basis function, Bessel function, Green function, frequency spectrum, Conductivity, $\lambda = \frac{C}{F}$, dimension $\frac{\lambda_g}{\epsilon_r}$, wave function, miniaturization/de-miniaturization, group delay, loss tangent, characteristics impedance ($\eta = \sqrt{\frac{\mu_0}{\epsilon_0}} = 377\Omega$), permittivity free space $= 8.854 \times 10^{-12} / \frac{F}{m}$, permeability free space $= 4\pi \times 10^{-7}\ H/_m$, short magnetic dipoles, radiation resistance, image effect, dominant mode, absorption, etc. Antenna radiating properties are mostly known as reactive zones, near field zones and far field zones. These are given by $r < \lambda/2\pi$, and this relationship is the reactive zone, where r is distance in meters between source and destination, i.e., transmitter antenna and receiving antennas. This reactive zone is not useful for nay communication. When, $\frac{2D^2}{\lambda} <r> \frac{\lambda}{2}$, the range is known as near field zone. This is also called as the Fresnel zone and is used for Wi-Fi and Bluetooth communication. The third relationship, $r > \frac{2D^2}{\lambda}$, is known as the far field zone, also well known by the name Franhaufer zone. The most useful range for radar and communication systems is $r > 10\lambda$. The indoor measurements are carried out in an EM controlled environment. The isolation level must be better than—50 dB, between transmitted and reflected signals. The main objective of indoor range is to reduce unwanted reflections by making use of RF absorbers at metal surface and focal lens at source antenna. Sometimes two lenses can be placed in cascade to achieve a DUT oriented focused beam. WH emersion described a detailed study of anechoic chamber design and development using mathematical formulations of fast Fourier transform for conversion of near field measurement into far field. Planar coordinates can be transformed into spherical ones and vice versa. In an anechoic chamber, a strong correlation is obtained between absorbers, reflectivity and quiet zone.

The energy (EM wave) is mostly incident on DUT from source antenna. The fraction energy travels toward the chamber walls, ceiling and floor. RF absorbers are pasted and clamped on the walls. The EM wave may incident at a normal or oblique angle on the absorbers. The wave incident at normal travels through carbon coated absorbers, gets converted into heat and radiates out in the space through absorbing material as dissipated heat. Thus, unwanted reflections are absorbed. There can be two technical reasons to handle these reflections. This can be either through resonance or resistivity offered by the absorber. These are incident electromagnetic wave frequency and impedance dependent. They are also power dependent. Absorbers form a crucial part in any anechoic chamber. Absorption or attenuation of RF energy takes place due to the material characteristics of the absorber. They are made from urethane embedded with spray paint or carbon coating. RAM (radiation absorbing material) is used as an absorber. Less reflectivity in dB may result in a large quiet zone. Thus, RF absorbers help to minimize secondary scattered reflections $\lambda / 4$. Size absorbers are used *in* the chamber. The tip of pyramid shaped absorbers is matched with space impedance.

These test facilities are classified as indoor/outdoor, such as: (a) EMI/EMC/ESD/ Immunity/CE/CS test facility, (b) near field test facility, (c) compact ranges test facility and (d) far field test facility.

The technical specifications and applications classify the type of facility developed, i.e., 5 or 10 meter range. These are most popular test facilities used worldwide. The size of the quiet zone in indoor test facilities defines the anechoic chamber usage. Normal angle of incidence and oblique angle of incidence may define quantity of reflections inside the chamber. Absorber made from meta material can further help reduce reflections. This may result in better reflectivity and an enlarged quiet zone. The compact test range was invented by Richard Johnson and Dorech Hess in 1969. Dominant parameters are r/λ and D/λ, where D is size of DUT and r is distance between transmitter and receiver antennas. A rectangular room is built with metallic walls. All the inner surfaces of the shielded room are then covered with foam absorbers. These are wall mounted with brackets and pasted with glue. A transmit antenna radiates out energy at a specified frequency and power of our interest. The same is received by the receive antenna as a direct beam (line of sight). It is observed that there are few unwanted EM signal hits at walls and absorbers at normal incidences and oblique incidences. If incidence waves are at normal, they are absorbed by absorbers. Hits at oblique angles may get reflected. This causes interference or distortions to the receive antenna. Figure 11.6 represents the layout of an anechoic chamber.

A transmitter antenna connected with signal generator, rotor/turn table/positioner with pay load capacity and rotating platform for DUT, VNA and control unit is interfaced with a plotter as shown in Figure 11.2.

The quiet zone is known as a reflection free space. The absorption power can be calculated using a set of source antenna, absorber and receive antenna. The source is oriented toward the RF absorber placed on the wall. In this set up, receive antenna is connected to receive reflected power. The transmitter antenna is connected with a signal generator, and the receive antenna is connected with VNA (vector network analyzer). Figure 11.7 shows absorbers of FSS, foam and hybrid types.

Reflectivity measurement is carried out using one signal generator (TXR) and one spectrum analyzer (RXR) along with absorbers tightly clamed on the wall

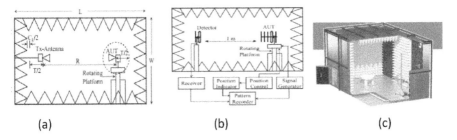

(a) (b) (c)

FIGURE 11.6 (a) Basic building block, (b) complete architecture, and (c) physical anechoic chamber with absorbers.

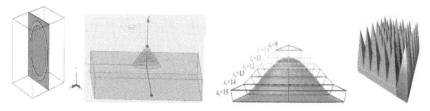

(a) FSS absorber model (b-c) Hybrid absorber model   (d) Pyramid Foam absorber

FIGURE 11.7   (a) FSS absorber model, (b and c) hybrid absorber model, and (d) pyramid foam absorber.

(foam and carbon coated, FSS [frequency selective surface] and hybrid RF absorbers). The parameters of antenna are measured using ray tracing or beam tracing methods in the anechoic chamber. Polyurethane carbon impregnated absorbers reduce internal reflections, thus providing measurements which are free from ambiguous extraneous signal interference and reflections or echo free region. This echo free zone is also known as the quiet zone. Thus, the RF controlled environment is achieved by means of anechoic chamber for EM radiation measurement. The quiet zone size will depend on $n\lambda$; where n is indices and λ is wave length. The quiet zone (reflection free volume inside anechoic chamber) key design item for any anechoic chamber is the RF absorbers. They are responsible for minimizing reflected energy by means of resonance or resistivity. These are made of complex permittivity or permeability, thus they give rise to lossy components. These absorbers are perfectly matched with free space impedance 377Ω. The tips of the absorbers allow incidence RF energy to flow through their entire length only when wave falls at a normal incidence angle. RF energy gets converted into thermal energy and dissipated as heat into surrounding air. Electrical length and thickness of the absorber will be responsible for the amount of absorption power. EMI/EMC chambers are designed for low frequency, and antenna measurement chambers are designed for high frequency measurements. The anechoic chamber with absorbers installed on walls, floor and ceiling reduce reflectivity.

These three types of test facilities are named as: (a) compact range test facilities, (b) near field test facilities and (c) far field test facilities or outdoor test facilities. These are shown in Figure 11.8a–d. DUT is the device under test. DRA is tested for radiation patterns inside the chamber.

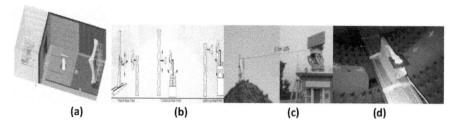

(a)   (b)   (c)   (d)

FIGURE 11.8   (a) Compact range test facility, (b) near field test facility, (c) far field test facility, and (d) DUT.

11.3 MATHEMATICAL FORMULATIONS

Currents at surface of DUT are transformed using Fourier transform by an inbuilt software for far field radiation patterns. The mathematical formulations from equations (11.1)–(11.4) present the EM wave solution by transforming current density into corresponding radiated fields inside the anechoic chamber.

Maxwell's equation for current density and radiated fields mathematical solution is presented in equations (11.1)–(11.4) as below:

$$\left(\nabla^2 + k^2\right)E_\perp = j\omega\mu\,\underline{J}_\perp;\left(\nabla^2 + k^2\right)H = -\nabla \times \underline{J}; j\omega c\; E_\perp = \left(\nabla \times H\right)\perp - \underline{J}_\perp;$$

$$\left(\nabla^2 + k^2\right)H_\perp = -\left(\nabla \times j\right)\perp;$$

$$\underline{J}_\perp = Re\left[\sum_{m,n,p=1}^{\infty} C\left(m,n,p\right)\left(\frac{m^2}{a^2} + \frac{n^2}{b^2} + \frac{p^2}{d^2}\right)\pi^2 \frac{\omega^2\left(mnp\right)}{\left(\frac{p\pi z}{d}\right)} e^{-j\omega\left(mnp\right)}\right]$$

$$\cos\left(\frac{m\pi x}{a}\right)\cos\left(\frac{n\pi y}{b}\right)\sin\left(\frac{p\pi z}{d}\right); \tag{11.1}$$

The radiated far fields:

$$H_z\left(x,y,z,t\right) = Re\left[\sum_{m,n,p=1}^{\infty} C\left(m,n,p\right)e^{j\omega\left(mnp\right)t}\cos\left(\frac{m\pi x}{a}\right)\cos\left(\frac{n\pi y}{b}\right)\sin\left(\frac{p\pi z}{d}\right)\right]; \tag{11.2}$$

$$E_{\text{farfield}} = -jk\frac{e^{-jkr}}{2\pi r}\hat{r}\int_b\int_a \hat{n}\times E e^{jkr'}\, dx'\, dy' \tag{11.3}$$

$$\text{Gain} = \sqrt{\frac{1}{2}\left[20\log_{10}\left(\frac{4\pi R}{\lambda}\right) + 10\log_{10}\left(\frac{P_r}{P_t}\right)\right]}; \tag{11.4}$$

11.4 HFSS SIMULATED RESULTS

Reflectivity measurements with hybrid type absorbers (hybrid absorber enclosed by multi dielectrics surfaces to provide cloaking effect) are used inside anechoic chambers. The field strength at any point inside the anechoic chamber can be assessed by measuring field strength of our choice and oprating frequency bandwidth. For this we make use of known parameters for antennas and source. By method of comparison, reflectivity can be assessed. The absorber characteristics can also be studied in a similar manner. Comparative studies of different approch have been carried out using HFSS and CST software. Optimzed results have been obtained and reported in

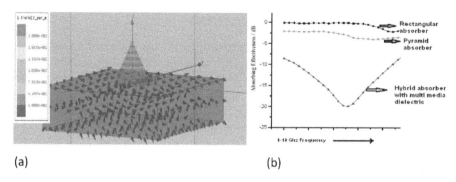

(a)   (b)

FIGURE 11.9   (a) E-Fields plot of hybrid absorber and (b) relectivity simulated plot.

this paper. Figure 11.9a and b is based on HFSS and CST softwares. These are simulated results showing fields and reflectivity. It has been seen that improved reflectivity was obtained when simulated with hybrid absorbers.

11.5   CONCLUSION

Reflectivity is an important parameter inside an anechoic chamber. For accurate measurements of antenna radiation pattern, reflections inside the chamber must be minimal. These can be minimized by making appropriate choice of proposed hybrid absorbers. Absorbers can provide better reflectivity at oblique incident angles. Reflections have been a major concern of anechoic chamber designers. These can be best minimized by making use of size, shape and hybrid absorbers. New embedded multi dielectric layers reduce reflection in the chamber. They have given cloaking effects and are able to reroute the e.m. waves. The incident wave gets converted into a normal incidence wave. Hence, reflections are reduced. The proposed work has better reflectivity. The proposed structure of hybrid absorber is unique and novel. Hybrid pyramid absorbers have further reduced the reflectivity.

The absorption coefficient can be measured with one port as well as with two port analysis using VNA. The absorption coefficient depends on material, shape and size of absorber used inside the anechoic chamber. The size of the absorber is wavelength specific. These absorbers are made from magnetic materials. Internet on things and 5G are current trends in data and voice communication. Wireless communication needs faster bit rates transfer. Antenna is an integral part of wireless communication systems and must keep pace with communication bandwidth requirements. These chambers are classified by EMI/EMC/ESD (electromagnetic interference/electromagnetic compatibility/electrostatic discharge) and radiation pattern measurements. EMI/EMC/ESD measurements are low frequency tests. Thus, reflections are minimized. The mathematical background of the problem is developed in the formulation part.

REFERENCES

1. F.-W. Trautnitz, EMC Absorbers through the Years with Respect to the New Site VSWR Validation Procedure in the Frequency Range from 1 to 18 GHz: A Practical Approach, *IEEE 2007 International Symposium on Electromagnetic Compatibility*, Honolulu, HI, July 9–13, 2007, pp. 1–6.
2. I. J. Gupta, "Compact Range Reflector Edge Treatment Impact on Antenna and Scattering Measurements," *IEEE AP-Magazine*, vol. 46, pp. 20–29, June 2004.
3. Design and engineering of a large shielded semi-anechoic chamber meeting the volumetric NSA requirements at 3 and 10 m transmission length Franconia, Germany.
4. S. Ghosh and K. V. Srivastava, "Polarization-Insensitive Single and Broadband Switchable Absorber/Reflector and its realization using a Novel Biasing Technique," *IEEE Trans.*, vol. 64, no. 8, pp. 3665–3670, 2016.
5. L. H. Hemming, *Electromagnetic Anechoic Chambers: A Fundamental Design and Specification Guide*. Wiley-IEEE Press, June 2002.
6. M/s Franconia Germany, M/s Orbit France, M/s Microwave vision Group, M/s JV Micronics, Emersion and cummings and M/s Sahajanand.
7. S. Ramo, J. R. Whinnery, and T. Van Duzer, *Field and Waves in Communications Electronic*, New York: John Wiley & Sons, 1965.
8. C. L. Holloway, R. R. DeLyser, R. F. German, P. McKenna, and M. Kanda, "Comparison of Electromagnetic Absorber Used in Anechoic and Semi-Anechoic Chambers for Emissions and Immunity Testing of Digital Devices," *IEEE Trans. Electromagn. Compat.*, vol. 39, no. 1, pp. 33–47, 1997.
9. J. D. Kraus, *Electromagnetics*, 4th ed. New York: McGraw-Hill, 1992. Broad-Band EMI Suppression Absorber.
10. M. B. Amin and J. R. James, "Techniques for utilization of Hexagonal Ferrites in Radar Absorbers Broad Band Planar Coating," *J. Inst. Electron. Radio Eng.*, vol. 51, no. 5, pp. 209–218, 1981.
11. C. L. Holloway and E. F. Kuester, "Modeling Semi-Anechoic Electromagnetic Measurement Chambers," *IEEE Trans. EMC*, vol. 38, no. 1, pp. 79–84, 1996.
12. B. K. Chung, C. H. Teh, and H. T. Chuah, "Modeling of Anechoic Chamber Using Abeam-Tracing Technique," *Prog. Electromagn. Res.*, vol. 49, pp. 23–38, 2004.
13. H. Togawa, K. Hatakeyama, and K. Yamauchi, "Reflectivity Measurements in Anechoic Chambers in the Microwave to Millimeter Range," *IEEE Trans. Electromagn. Compat.*, vol. 47, no. 2, pp. 312–319, 2005.
14. V. Trainotti, "Electromagnetic Compatibility (EMC) Antenna Gain and Factor" *IEEE Trans. Electromagn. Compat.*, vol. 59, no. 4, pp. 1006–1015.
15. A. Farahbakhsh, M. Khalaj-Amirhosseini, and H. Oraizi, "Ellipsoid Anechoic Chamber for Radiation Pattern Measurements" *2015 IEEE 15th Mediterranean Microwave Symposium (MMS)*.
16. P. Piasecki and J. Strycharz, "Measurement of an Omnidirectional Antenna Pattern in an Anechoic Chamber and an Office Room with and without Time Domain Signal Processing." *2015 Signal Processing Symposium (SPSympo)*, IEEE.
17. H. Nornikman, F. Malek, M. Ahmed, F. H. Wee, P. J. Sohand, and A. A. H. Azremi, "Setup and Results of Pyramidal Microwave Absorbers Using Rice Husks" *Prog. Electromagn. Res.*, vol. 111, pp. 141–161, 2011.
18. A. Farahbakhsh and M. Khalaj-Amirhosseini, "Design of Non uniform Metallic Anechoic Chamber for Radiation Pattern Measurement," *Prog. Electromagn. Res.*, vol. 58, pp. 65–72, 2017.

12 Vehicular Smart Antenna

12.1 INTRODUCTION

Wireless sensors have made human life comfortable. Smart vehicles are equipped with facilities of all required infotainment and entertainment features. Smart vehicles can communicate wirelessly and can sense surroundings even when visibility is zero due to fog. Sensors are also used in vehicles to communicate with infrastructure for collection and monitoring of data such as display of outside temperature, traffic status of routes in real time basis and status of passengers on board using biomedical wearable sensors that can be remotely monitored [1]. These vehicles need wireless communication that is independent of vehicle position and speed [2]. The vehicles' capability shall be rated as serve capability, i.e., entertainment and infotainment [3]. Smart vehicles in smart cities are an excellent amalgamation for futuristic living [4]. Antennas are an important part in all vehicles to make wireless communication between two nodes possible. Hence, vehicle to vehicle, vehicle to infrastructure, vehicle to satellite and vice versa all need wireless communication for smart mobility [5]. Antenna plays a vital role to make all vehicles autonomous or smart. These vehicles are facilitated with surveillance needs such as Global Positioning System (GPS), cellular services, i.e., Global System for Mobile Communications (GSM), AM/FM radio, Digital Audio Broadcasting (DAB), Satellite Digital Audio Radio System (SDARS), remote key entry system (RKES), VHF TV, tire pressure monitoring system (TPMS), collision avoidance sensor, electronic toll collection system (ETCS), Wi-Max, Wi-Fi, WLAN and Bluetooth services [6]. Soon 4 and 5G communication features will become the need of hour in all smart vehicles. This will enable them to take real time decisions [7]. Signal quality plays crucial role. All wireless equipment's working depends on signal strength [8]. Use of sensors can impart a three dimensional view of the road few meters ahead when the vehicle is on the move.

There used to be many antennas installed on the roof of a vehicle to serve all above mentioned requirements due to different operating frequency bands [9]. There is a need for omnidirectional, directional and isotropic antennas to make smart vehicles. Navigational services need directional antennas. For broadcast signal reception, there is a need for omnidirectional antennas [10]. Sensors for safety and security are in built with advanced features such as remote key entry (RKE) [11]. Monopole antenna, dipole antenna, helical, printed inverted F type antenna (PIFA), compact inverted type antenna (CIFA), meander line patch and dielectric resonator antenna (DRA) have been proposed by many authors for unidirectional, directional and isotropic signal reception in automotive applications [12]. Frequencies of 0.1–500 MHz are received with printed on glass antenna. These are meander line patch antenna for 500–2500 MHz frequency and 24–77 GHz is received with mm Wave radar antennas in vehicles.

Presently 150–1600 KHz band is assigned for AM radio and 80–108 MHz is meant for FM radio [13]. For DAB, 217.5–250 MHz band is allocated, and 174–240 MHz,

470–840 MHz are reserved for VHF TV [14]. For remote key entry and tire pressure monitoring, 315, 413, and 434 MHz are allotted. Omnidirectional type antenna for RKES and isotropic type antenna for TPMS is used [15]. Frequencies of 900, 1800, and 2300 MHz are used for GSM; 1575–1602 MHz is used for GPS for surveillance and 2.1–2.3 GHz is used for SDAR. Wi-Fi and Bluetooth use 2.4–2.5 GHz; 2.3, 2.5, and 3.5 GHz is used for Wi-Max and 2.4 and 5.8 GHz is used for wireless local area network (WLAN) [16]. Electronic toll collection (ETC) is allotted 900 MHz and 5.8 GHz [17,18]. Short range radar (SRR) and long range radar (LRR) are used as collision avoidance radars, operating in the 22–77 GHz frequency range. Vehicle to vehicle communication operates at 5.9 GHz. Zigbee *Tx/Rx* operate at 4.9–6 GHz frequency spectrum [19], and 4.4–4.9 GHz is used for international mobile telecommunication (IMT) [20].

Polarization and gain of each type of antenna have different roles in communication networks. Polarization can be used to extend diversity in communication systems and range used by increased gain. Isolation, frequency ratio and power ratio are a few important parameters for compact and multiband antennas [21,22]. SRR and LRR are radar mm wave antennas used to add safety features into the vehicle. With the help of LRR, it activates the breaking system before collision [23,24]. TPMS sensors monitor tire pressure of all four wheels in an instant and show it in the display. Temperature sensor monitors inside and outside temperature of the vehicle. If a wireless embedded jacket is worn by the passenger, all biomedical parameters such as pulse rate/heart rate/blood pressure/body temperature, etc. can be monitored remotely. Infotainment features such as communication with the environment while moving can update status about restaurants, airports, cinemas, shopping complexes, traffic, train schedules and others on a real time basis. Thus, self-driving cars become a possibility with embedded wireless sensors.

M/s AGC has developed integrated glass antennas for receiving mobile services in vehicles. M/s Zhejiang JC Antenna Co. Ltd. M/s SIGNAXO, M/s Jiashan Jinchang Electron Co. Ltd, M/s JC Antenna HK Co Ltd and M/S G-Antetech Industrial Co., Ltd. Information about congestion, traffic jams, etc. can be known ahead to an intelligent vehicle. The dashboard, top of front glass, rooftop and rear side glass are the best places for installation of automotive antennas. Light detection and ranging (LIDAR) is a pulsed laser light system used to scan and capture images of surroundings. This can provide the motorist a three dimensional view of the road ahead. These radars generally operate in mm wave range of frequencies. M/s Vellodyne is manufacturer for LIDAR. Laser sensors are used in vehicles to scan 360 degree view of the vehicle environment i.e., front side and rear side view. The vehicle can see up to 100 meters even in fog when visibility is low. Thus, real time object recognition when embedded with navigation features makes the car intelligent and self-driving.

This antennas chapter discusses minimizing the number of antennas in vehicle and reduction in antenna size. This shall be used in advanced smart vehicles. An amalgamation of meander line and dielectric resonator antenna can be the best option for application to automotive vehicles with minimal space requirements in the vehicle. AM/FM/DAB/VHF TV/RKE antenna services can be integrated in car windshield glass, and all other services such as SDAR, TPM, GPS, GSM, ETC, Wi-Fi, Bluetooth, WLAN and SRR, SPRR collision avoidance can be extended with the shark fin type antenna. A compact plastic box, i.e., redome, can house a single MIMO dielectric resonator antenna. All these services of infotainment and entertainment thus become possible in the vehicle.

Defected ground plane structure (DGS) can be used to reduce mutual coupling between antennas. Higher order modes can be been used for antenna compactness.

The articles in this chapter discuss designs of long range radar car sensors at 24 GHz with prototype model. Results of simulations and measurements are also presented along with discussion of futuristic quantum antenna which can be used for wireless charging of vehicles. Review on various types of vehicle antennas have been tabulated in Table 12.1. NDRA is a futuristic candidate to be used for long range scanning for surveillance and navigation during bad weather, including fog. At terahertz frequencies, NDRA with higher order modes concepts can be used to provide the facility of beams forming an electronic runway for aircraft landing and takeoff in bad weather conditions. This antenna is very effective when visibility is poor. This can replace the present CAT III system, as it has limited capability. The frequency spectrum used for smart vehicle communications is given in Table 12.1. It has also been studied that by varying the axial ratio, the beam width of the antenna can also be controlled.

TABLE 12.1
Frequency Spectrum Used for Vehicle Communication

Service	Typical Frequency
AM radio	Approximately 1 MHz
FM radio	88–108 MHz
In-vehicle VHF-TV	50–400 MHz
Digital Audio Broadcasting (DAB)	100–400 MHz
Remote keyless entry (RKE)	315/413/434 MHz
Tire pressure monitoring system (TPMS)	315/413/434 MHz
Cellular phone, GSM	850, 900, 1800, 1900 MHz, (2100 MHz)
IMT (International mobile telephony)	2300–2400 MHz
	2700–2900 MHz
	3400–4200 MHz
	4400–4900 MHz
Satellite navigation (GPS)	1.575 GHz
WLAN	2400–2485 MHz
	5150–5350 MHz
	5725–5825 MHz
Satellite Digital Audio Radio Service (SDARS)	2.3 GHz
Wi-Fi	2.4/5.8 GHz
Bluetooth	2.4 GHz
Wi-Max	2.3/2.5/3.5 GHz
	2500–2690 MHz
	3400–3690 MHz
	5250–5850 MHz
Electronic toll collection (ETC)	5.8 GHz (or 900 MHz)
V2V (vehicle to vehicle)	5.9 GHz
Collision avoidance radar (LRR & SRR)	24 and 77 GHz
5G communication	24.5–27.5 GHz and 1–6 GHz
NDRA	192 THz

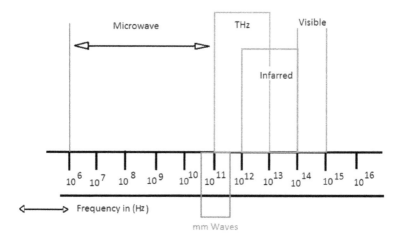

FIGURE 12.1   Frequencies used in microwave communication, infrared and quantum communication.

12.2   ANTENNA DESIGN

The proposed design is an amalgamation of meander line patch antenna for AM/FM/VHF TV and DAB services embedded in FM chip antenna. Then, meander line antennas and GPS/GSM/ETC/Wi-Fi/Bluetooth/SDAR, and SRR with LRR on other corner of PCB are embedded with dielectric resonator antennas at mm wave. Microwave frequencies using fundamental and higher order mode concepts have been implemented. There should be proper isolation between working antennas. Minimum frequency ratio and desired power ratio in each multi band antenna are essential requirements. The mm wave SRR and LRR can be installed easily in the front bumper of the vehicle to detect any object approaching close by. ETCS antenna is in communication with beacon antenna inline of sight (LOS) mode installed at a toll gate. ETCS is highly directive in nature with narrow beam width. Sensors are integrated all around the vehicles for safety and security purposes. TPMS sensors communicate with embedded vehicles in Bluetooth mode at 2.4 GHz. GPS antenna is designed to communicate with satellite GPS at 1575 MHz as directive antennas. The FM/AM/DAB/VHF TV operate in an omnidirectional pattern. Automated vehicles have space diversity due to multiple input multiple output (MIMO) antenna and pick up signals from different directions. In MIMO architecture, good isolation <-25 dB between two radiation operating bands and diversity parameters. Reflection coefficient is better $(S_{11}) > 10$ dB; ECC (envelop correlation coefficient) to assess coupling fields between two antennas should be minimum. They can be determined using the far field method.

Glass integrated antenna, meander line patch, dielectric resonator antenna and active antenna (as shark fin antenna) are many in one antenna, a new type of antenna to facilitate all services required for self-driving vehicles. They are also self-charging vehicles. LRR and SRR antenna based on DRAs can be integrated in the front side of the vehicle with beam width control features. Figures 12.1 through 12.6

are used to validate working of long range radar at 24 GHz, very useful for scanning. Figures 12.7 through 12.11 illustrate working and radiation pattern validation of NDRA at 192 THz.

12.2.1 STRUCTURE: LRR AUTOMOTIVE RDRA

Aperture coupled rectangular dielectric resonator antenna has been modeled for LRR (long range radar) vehicular applications.

Boundary conditions are shown as PEC for electric walls and PMC for magnetic walls; further, $n \times E = 0$ and $n. H = 0$ is used for PEC walls; $n \times H = 0$ and, $n. E = 0$ is used for PMC walls. As per the law of conservation of energy, integration of $|E^2|$ in cavity volume = integration of $|H^2|$ in the same cavity volume. Micro strip line design equations and transcendental equations are given below:

Micro strip line calculation:

$$A = \frac{Z_{0m}}{60}\left\{\frac{\varepsilon_r + 1}{2}\right\}^{1/2} + \frac{\varepsilon_r - 1}{\varepsilon_r + 1}\left[0.23 + \frac{0.11}{\varepsilon_r}\right] \qquad (12.1)$$

$$W/h = \frac{8\exp(A)}{\exp(2A) - 2}; \qquad (12.2)$$

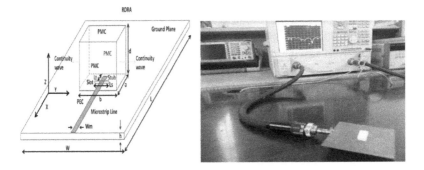

FIGURE 12.2   V2V (vehicle to vehicle) antenna (with ZIRCAR DR at 5.9 GHz).

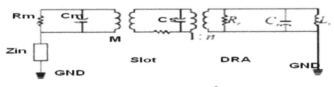

FIGURE 12.3   Equivalent circuit of RDRA.

Effective permittivity

$$\varepsilon_{reff} = \frac{\varepsilon_r + 1}{2} + \frac{\varepsilon_r - 1}{2} \left[1 + 12 \frac{h}{W} \right]^{-\frac{1}{2}} \quad (12.3)$$

Transcendental equation for resonant frequency calculation:

$$\varepsilon_r \, ko^2 = kx^2 + ky^2 + kz^2; \quad (12.4)$$

$$k = \omega\sqrt{\mu\varepsilon}$$

where:

$$kx = \frac{m\pi}{a}; \quad ky = \frac{n\pi}{b}; \quad kz = \frac{p\pi}{d}; \quad (12.5)$$

$$k_z \tan\left(k_z d/2\right) = \sqrt{\left(\varepsilon_r - 1\right) k_0^2 - k_z^2}$$

DR of dimension rectangular $12 \times 12 \times 12.4$ mm at the resonant frequencies $f_r = 24$ GHz, with two sides equal $(a = b)$, RDRA $(\varepsilon_r = 10)$. $L_{st} = 30$ mm, $W_{st} = 30$ mm, $W_{sl} = 1$ mm, $L_{sl} = 5$ mm RDRA has been fabricated and mounted on a FR4 $(\varepsilon_r = 4.4$ and $\tan\delta = 0{:}0019)$. Substrate thickness is 1.6 mm and micro strip feed line width was set to have 50 ohm and ground plane 30×30 mm^2. Higher mode (TE$^y_{117}$, based on aspect ratio and slot position) has been excited into RDRA. Y indicates that h field is propagating in the y direction, while excitation field has been applied orthogonal, hence along the x direction. An angular rectangular strip placed on top of DRA introduces circular polarization. LHCP/RHCP will depend upon which field is dominant. Undesired modes can be suppressed using defected ground structure concepts. Circular polarization eliminates the requirement of line of sight and multipath fading to establish communication. The ground dimensions are taken more than DRA as given below:

$Lg = L + 6\,h$; where, h is height of substrate, $Wg = W + 6\,h + 6\,h$;
$\lambda = \frac{c}{f}$; operating wavelength of antenna;

$$\text{permitivity } \epsilon r_{\text{eff}} = \left(\frac{\lambda_o}{\lambda_g} \right)^2;$$

Resonant frequency of DRA is calculated by the formulation given below:

$$\epsilon_r = \frac{\epsilon}{\epsilon_o};$$

$$\frac{\lambda}{2\pi} < r < \frac{2D^2}{\lambda}; \text{ Near field}$$

$$k \gg 1; \text{ far field } r > \frac{2D^2}{\lambda};$$

$$fr_{mnp} = \frac{c}{2\pi\sqrt{\varepsilon r}} \sqrt{\left(\frac{m\pi}{a}\right)^2 + \left(\frac{n\pi}{b}\right)^2 + \left(\frac{p\pi}{2d}\right)^2} \tag{12.6}$$

Slot length,

$$L_S = \frac{0.4\lambda_o}{\sqrt{\varepsilon_e}}$$

where λ_o is the wavelength and the effective permittivity is defined as:

$$\varepsilon_e = \frac{\varepsilon_r + \varepsilon_s}{2}$$

where ε_r and ε_s are the dielectric constant of the rectangular dielectric resonator and substrate, respectively.

Slot width,

$$W_s = 0.2L_s$$

And the stub length,

$$St = \frac{\lambda_g}{4} \text{ and } \lambda_g = \frac{\lambda_0}{\left(\varepsilon_{eff}\right)1/2} \tag{12.7}$$

where λ_0, λ_g are free space and guided wavelengths.

Circular polarization was achieved by placing rectangular strip on top of rectangular DRA at 45° to the aperture coupled slot.

12.3   WI-FI VEHICULAR DRA AT 3.6 GHZ

DRA for Wi-Fi has been designed and simulated. Results of S_{11}, E-fields, radiation patterns, etc., have been obtained and presented below:

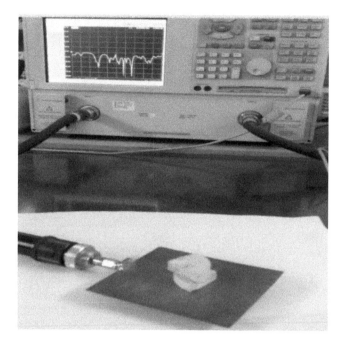

FIGURE 12.4   Electronic toll collection antenna (designed to operate at 5.8 GHz).

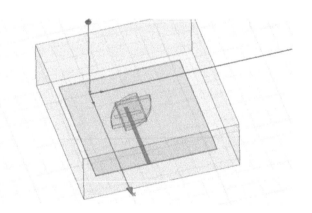

FIGURE 12.5   DRA HFSS model.

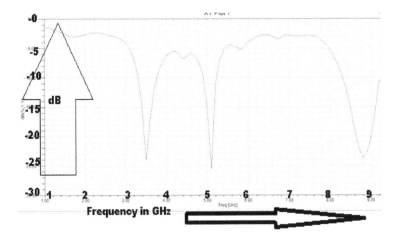

FIGURE 12.6   DRA S_{11} simulated.

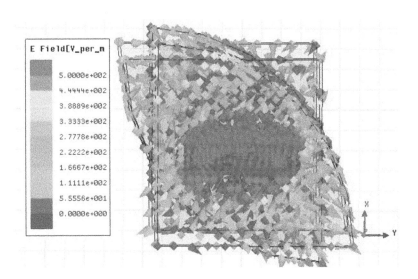

FIGURE 12.7   DRA E-fields.

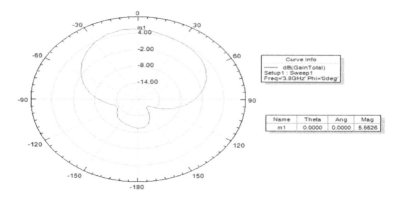

FIGURE 12.8 DRA radiation pattern.

12.4 MATHEMATICAL FORMULATIONS

Meander line patch antenna resonant frequency is calculated as $f_r = \frac{c}{2L\sqrt{\varepsilon_{eff}}}$; c is velocity of light, L length of antenna and ε_{eff} effective permittivity of substrate. Meander line antennas are used for RFID in electronic toll collection with elliptical polarization. Backscattering takes place due to mismatch between chip and antenna of RFID.

$$\text{Read range of RFID}, r = \frac{\lambda}{4\pi} \cdot \left[\frac{Pt.Gt.Gr.p.\tau}{Pth} \right]^{1/2};$$

p is polarization efficiency and τ is transmission coefficient.

$Zc = Rc + jXc$ and $Za = Ra + jXa$: are chip and antenna impedances, respectively. They are functions of frequency along with range. Hence, reader, chip, tag and range are important parameters in RFID.

MIMO envelop correlation coefficient (ECC) defines coupling fields to nearby radiators as:

$$\text{ECC}_f = \frac{\int\int_{4\pi} \left[A_i(\theta,\phi) \times A_j(\theta,\phi) \right] d\omega}{\int\int_{4\pi} \left[A_i(\theta,\phi)^2 \times \int\int_{4\pi} A_j(\theta,\phi)^2 \right]}; \quad A_i(\theta,\phi) \text{ and } A_j(\theta,\phi) \text{ are far}$$

field patterns when i and j ports are excited. The same can be obtained using s parameters as:

$$\text{ECC}_\beta = \frac{\left| S_{11}\, S_{12} + S_{21}\, S_{22} \right|}{\left[\left(1 - \left(|S_{11}|^2 + |S_{21}|^2 \right) \right) \left(1 - \left(|S_{22}^2| + |S_{21}^2| \right) \right) \right]}; \quad S_{11},\, S_{22} \text{ are reflection}$$

coefficients and S_{12} and S_{21} are mutual coupling.

$$\text{Diversity Gain, DG} = 10\sqrt{1-(\text{ECC})^2} \tag{12.8}$$

The diversity parameters are ECC (envelope correlation coefficient), DG (directive gain), MEG (mean effective gain) and CCL (coupling coefficient or signal transfer rate). MEG is gain of antenna (MIMO) under certain defined environmental conditions. MIMO with ports is given below, only for representation purposes without HFSS analysis:

(DRA dimensions ($L_g = 80$ mm, $R = 10$ mm, $H = 10$ mm, $W_f = 1.6$ mm, $t_h = 1.67$ mm)

DRA (high frequency) with meander line (low frequency) as MIMO for FM at 88 MHz and Wi-Max at 3.7 GHz:

$$\text{MEG}_1 = 0.5\left[1 - |S_{11}|^2 - |S_{12}|^2\right]$$

$$\text{MEG}_2 = 0.5\left[1 - |S_{12}|^2 - |S_{22}|^2\right]$$

Power ratio, $k = |\text{MEG}_1 - \text{MEG}_2| < 3$ dB
    CCL is signal transfer rate:

$$\text{CCL} = -\log_2 \det\begin{bmatrix} (\beta_{11} & \beta_{12}) \\ (\beta_{21} & \beta_{22}) \end{bmatrix}$$

where:

$$\beta_{11} = 1 - \left(|S_{11}|^2 + |S_{12}|^2\right)$$

$$\beta_{22} = 1 - \left(|S_{22}|^2 + |S_{21}|^2\right)$$

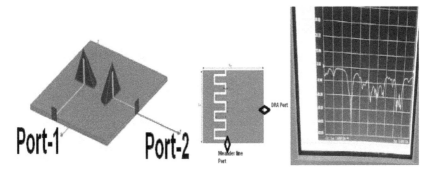

Port-1   **Port-2**

FIGURE 12.9 MIMO antenna with two ports.

$$\beta_{12} = -\left(S_{11}^{*}S_{12} + S_{21}^{*}S_{12}\right)$$

$$\beta_{21} = -\left(S_{22}S_{21} + S_{12}S_{21}\right)$$

Bits/second/hertz

12.5   RESULTS

Simulated and measured results at 24 GHz of LRR antenna have been obtained. Simulated results at 192 THz for quantum antenna (NDRA) have been recorded. Large range radar (LRR) automotive antenna was placed inside an anechoic (an echo free measurement set up) chamber for measurements. Simulated results of both types of antennas are presented in Figures 12.1 to 12.11. Figures 12.12 to 12.15 present the working of horn DRA. Most of the antenna parameters such as radiation pattern, reflection coefficient, gain, axial ratio, bandwidth, etc. have been presented in this chapter. The vehicle is embedded with collision avoidance ability with use of

FIGURE 12.10   Long range vehicle radar rectangular DRA with wave launch connector (inside anechoic chamber measurement).

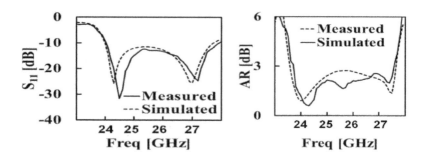

FIGURE 12.11   Simulated and measured S$_{11}$ and axial ratio of LRR.

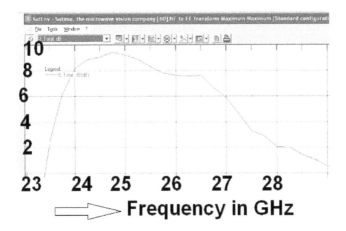

FIGURE 12.12 Measured gain 9.7 dBi max of LRR antenna at 24 GHz inside Satimo anechoic chamber.

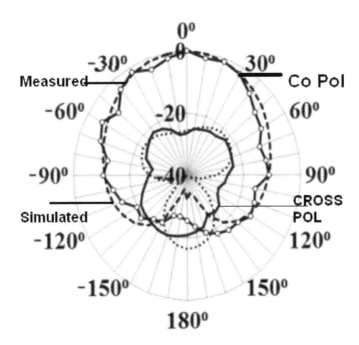

FIGURE 12.13 Simulated and measured radiation pattern (co and cross pol) at 24 GHz phi 0 and theta ±90° (xz plane).

LRR and SRR antennas. They can replace LASER sensors presently used in vehicles. LASER sensors are not human friendly. They are presently used for vehicle guidance systems. Radiation pattern, gain and axial ratio S_{11} have been measured and given below. They satisfy all the requirements for surveillance and guidance of vehicles (Figures 12.12 and 12.13).

12.6 HORN ANTENNA

12.6.1 DESIGN OF HORN ANTENNA AT 2.4 GHZ USING SUPERSTRATE OR PRS (PARTIALLY REFLECTED SURFACE)

The source antenna along with reflected antenna has partially reflected surface (PRS) above it and it is placed at a half wavelength ($\lambda/2$) distance. PRS and ground plane forms a cavity. Multiple reflections are observed to electromagnetic waves due to PRS. At a given frequency, constuctive obtructions or ineterferrences are observed to electromagnetic waves. They are partially transmitted through PRS. This phenonmenon helps to enhance directivity and beam width narrowing. Dierectivity is directly related to reflections and directly proportional to reflection coefficient magnitude (Figures 12.14–12.17).

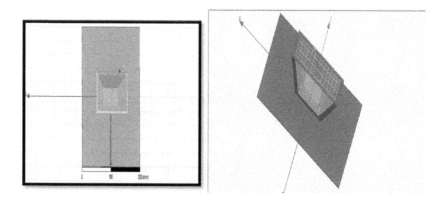

FIGURE 12.14 Horn antenna without superstrate and with superstrate.

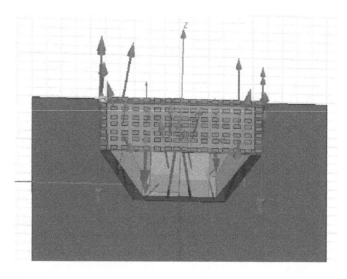

FIGURE 12.15 E-fields diagram.

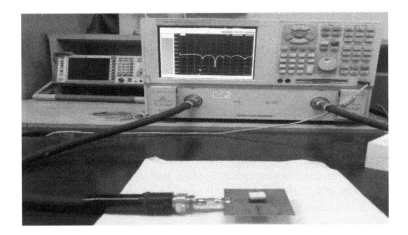

FIGURE 12.16   Reflection coefficient of horn antenna at 2.4 GHz (Wi-Fi).

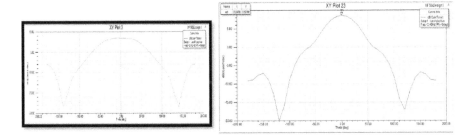

FIGURE 12.17   Beam width control in horn antenna using superstrate.

12.7   CONCLUSION AND FUTURE WORK

The outcome of this research is directly applied to self-driving and self-charging vehicles. The concept of NDRA array (nano dielectric resonator antenna array) on rooftops of smart vehicles is best suited for wireless charging of vehicles. Use of these embedded sensors shall be a boon to ease human life as an assistant to drivers. Different permittivity materials can be used with same size to realize sensors for different applications based on frequency response. Aesthetic design and less drag while moving are two important parameters. The place of installation of these sensors in automotive vehicles is also important. Integration of quantum antenna along with wind energy devices in futuristic energy efficient automotive vehicles may be the best idea for energy harvesting. These vehicles can have energy harvesting system such as solar cells or antenna on roof top for battery charging and fuel system for normal operation. Quantum antennas collect RF energy directly from satellites and deliver it to vehicles through a rectifier circuit. Similarly, wind energy is converted into DC voltage and is given to the vehicle. Thus, it can evolve as an efficient intelligent transportation system. In this chapter, the author proposed two antennas: one at 24 GHz for use as large range radar sensor (LRR) and another as nano DRA (NDRA) for energy extraction. They can directly communicate with satellites at 192 THz for energy harvesting. These NDRAs using higher order modes can replace LIDAR for long range scanning. The author purposes that the best use of NDRAs is in aircraft landing systems at airports. They can be good candidates for ILS at airports during foggy weather to counter low visibility. These multiple NDRAs placed in the center of the runway can also be used as multiple beacons operating at same frequency in sync with aircraft and can form an artificial electronic runway with excellent scanning to show a virtual runway on an aircraft screen to assist pilots during low visibility with naked eyes. Landing and takeoff during heavy fog conditions can be done with a 3D screen display of the runway in real time.

At present, radio based automatic instrument landing systems are being used. With this concept these systems will become radio integrated video based ILS. Integration of NDRA can improve Cat-III instrument landing systems significantly. Scanning of the runway becomes possibility with the use of quantum antennas. They can impart real time 3D visuals of runways during heavy fog. Use of NDRAs can open up pathways of easy landing features to the pilot by providing tunnel vision of runways with precision detection and ranging. These NDRAs can be very useful to build ad hoc networks to cover large distances with highly reconfigurable features.

Three types of substrates having different permittivity can have different beam width and can be used as LRR and SRR. 3D beam width is a radiation pattern solid angle that can be obtained with use of different materials of varying permittivity as composite substrates. NDRA can replace LIDAR (light detection and ranging) for possible extension of ranging and detection. NDRA operates at THz and is an excellent idea to develop intelligent runways at airports. This can result in smart runways to have a glance of artificial runways in bad weather or when fog is heavy during winter. Futuristic design of automotive vehicles can be extended to self-driving and self-charging features. Hence, integration of nano DRAs and sensors such as long range radar (LRR) and short range radar (SRR) can be used as collision avoidance and surveillance. Efficient and compact antennas for use of smart vehicles have been proposed. Application of MIMO antenna for vehicular antenna is better option.

The beam control feature can be added into an antenna. Thus, it becomes possible to control antenna beam width so that the same antenna can be used. Use of concave shape superstrates can enhance directivity and reduce beam width due to PRS (partially reflected surface). Operating at higher order modes further helps narrowing the beam width. The optimized beam width control can further be achieved by combination of PRS and operating at higher order modes. Self-reliance vehicles can be embedded with an array of nano DRAs on rooftops of vehicles for wireless energy harvesting.

REFERENCES

1. M. Liénard, P. Degauque, and P. Laly, "Long-Range Radar Sensor for Application in Railway Tunnels," *IEEE Trans. Veh. Technol.*, vol. 53, no. 3, pp. 705–715, 2004.
2. O.-Y. Kwon, R. Song, and B.-S. Kim, "A Fully Integrated Shark-Fin Antenna for MIMO-LTE, GPS, WLAN, and WAVE Applications," *IEEE Antenn. Wirel. Propag. Lett.*, vol. 17, no. 4, pp. 600–603, 2018.

3. S. Gotral, G. Varshney, R. S. Yaduvanshi, and V. S. Pandey, "Dual-Band Circular Polarization Generation Technique with the Miniaturization of a Rectangular Dielectric Resonator Antenna," *IET Microw. Antenn. Propag.*, vol. 13, no. 10, pp. 1742–1748, 2019.

4. N. Guan, H. Tayama, M. Ueyama, Y. Yasutaka, and H. Chiba, "A Roof Automobile Module for LTE-MIMO," *APS Topical Conference on Antennas and Propagation in Wireless Communications (APWC)*, pp. 387–391, IEEE, Torino, Italy, 2015.

5. W. Menzel and A. Moebius, "Antenna Concepts for Millimeter-Wave Automotive Radar Sensors," *Proc. IEEE*, vol. 100, no. 7, pp. 2372–2379, 2012.

6. D. Inserra, W. Hu, and G. Wen, "Antenna Array Synthesis for RFID-based Electronic Toll Collection," *IEEE Trans. Antenn. Propag.*, vol. 66, no. 9, pp. 4596–605, 2018.

7. R. S. Yaduvanshi and H. Parthasarathy, "Rectangular Dielectric Resonator Antenna Theory and Design," Springer, New Delhi, India, 2016.

8. G. Byun, Y. G. Noh, I. M. Park, and H. S. Choo, "Design of Rear Glass-Integrated Antennas with Vertical Line Optimization for FM Radio Reception," *Int. J. Auto. Technol.*, vol. 16, no. 4, pp. 629–634, 2015.

9. Q. Wang, H.-N. Dai, Z. Zheng, M. Imran, and A. V. Vasilakos, "On Connectivity of Wireless Sensor Networks with Directional Antennas," *Sensors*, vol. 17, no. 1, p. 134, 2017.

10. O.-Y. Kwon, R. Song, and B.-S. Kim, "A Fully Integrated Shark-Fin Antenna for MIMO-LTE, GPS, WLAN, and WAVE Applications," *IEEE Antenn.Wirel. Propag. Lett.*, vol. 17, no. 4, pp. 1536–1225, 2018.

11. K. Gong, X. H. Hu, P. Hu, B. J. Deng, and Y. C. Tu, "A Series-Fed Linear Substrate-Integrated Dielectric Resonator Antenna Array for Millimeter-Wave Applications" *Int. J. Antenn. Propag.*, vol. 2018, Article ID 9672790, 6 p.

12. B. Sain, C. Meier, and T. Zentgraf, "Nonlinear Optics in All-Dielectric Nanoantennas and Metasurfaces," *Adv. Photonics*, vol. 1, no. 2, p. 024002, 2019.

13. P. Rajalakshmi and N. Gunavathi, "Compact Complementary Folded Triangle Split Ring Resonator Triband Mobile Handset Planar Antenna for Voice and Wi-Fi Applications," *Prog. Electromagn. Res.*, vol. 91, pp. 253–264, 2019.

14. G. Clasen and R. J. Langley, "Meshed Patch Antenna Integrated into Car Windscreen," *Electron. Lett.*, vol. 36, no. 9, pp. 781–782, 2000.

15. Y.-M. Pan, K. W. Leung, and K.-M. Luk, "Design of the Millimeter-Wave Rectangular Dielectric Resonator Antenna Using a Higher-Order Mode," *IEEE Trans. Antennas Propag.*, vol. 59, no. 8, pp. 2780–2788, 2011.

16. G. Marrocco, "Gain-Optimized Self-Resonant Meander Line Antennas for RFID Applications" *IEEE Antennas Wirel. Propag. Lett.*, vol. 2, pp. 302–305, 2003.

17. G. Bakshi, R. S. Yaduvanshi, and A. Vaish, "Sapphire Stacked Rectangular Dielectric Resonator Aperture Coupled Antenna For C-Band Applications," *Wireless Pers. Commun.*, vol. 108, no. 2, pp. 895–905, 2019.

18. K. S. Sultan, H. H. Abdullah, and E. A. Abdallah, "Dielectric Resonator Antenna with AMC for Long Range Automotive Radar Applications at 77 GHz," *2018 IEEE International Symposium on Antennas and Propagation & USNC/URSI National Radio Science Meeting*, Boston, MA, 2018.

19. Q. Wu, Y. Zhou, and S. Guo, "An L-Sleeve L-Monopole Antenna Fitting a Shark-fin Module for Vehicular LTE, WLAN and car-to-car Communications," *IEEE Trans. Veh. Technol.*, vol. 67, no. 8, pp. 7170–7180, 2108.

20. P. Bartwal, "Design of Compact Multi-Band Meander-Line Antenna for Global Positioning System/Wireless Local Area Network/Worldwide Interoperability for Microwave Access Band Applications in Laptops/Tablets," *IET Microw. Antennas Propag.*, vol. 10, no. 15, pp. 1618–1624, 2016.

21. L. Low, R. Langley, R. Breden, and P. Callaghan, "Hidden Automotive Antenna Performance and Simulation," *IEEE Trans. Antennas Propag.*, vol. 54, no. 12, pp. 3707–3712, 2006.

22. R. Leelaratne and R. Langley, "Multiband PIFA Vehicle Telematics Antennas," *IEEE Trans. Veh. Technol.*, vol. 54, no. 2, pp. 477–485, 2005.

23. A. Ahmad Khan, R. Khan, S. Aqeel, J. Ur Rehman Kazim, J. Saleem, and M. K. Owais, "Dual-Band MIMO Rectangular Dielectric Resonator Antenna With High Port Isolation For LTE Applications," *Microw. Opt. Techn. Lett.*, vol. 59, no. 1, pp. 44–49, 2017.

24. A. Scannavini, L. J. Foged, T. Bolin, K. Zhao, S. He, and M. Gustafsson, "Radiation Performance Analysis of 28 GHz Antennas Integrated in 5G Mobile Terminal Housing," *IEEE Access*, vol. 6, pp. 48088–48101, 2018.

13 DRA as RLC Circuit with Resonant Modes

13.1 INTRODUCTION: DRA IS REPRESENTED AS RLC CIRCUIT

DRA has fundamental and higher order modes, they can be represented by RLC circuits [1–9].

A solution of general R, L, C, series circuit for current i (t), and resonant frequency (f_r) is shown in Figure 13.1.

Impedance (Z) power P and bandwidth (BW) are shown in Figure 13.2.

$$Z = R + j\omega L - \frac{j}{\omega C}$$

The circuit will be an integro-differential equation for the current $i(t)$ given below.

$$Ri(t) + L\frac{di(t)}{dt} + V_c(0^+) + \frac{1}{c}\int_0^t i\, dt = V_s \qquad (13.1)$$

Taking Laplace transforms of Equation (13.1), we get:

$$RI(s) + L\left\{ sI(s) - i_2(0^+) \right\} + \left\{ \frac{V_c(0^+)}{s} + \frac{I(s)}{sC} \right\} = \frac{V}{s}$$

Assume that there is no stored energy in the circuit at $t = 0$, hence $i(t0^+) = 0$ and $V_c(0^+) = 0$.

$$\therefore \left(R + sL + \frac{1}{sC} \right) I(s) = \frac{V}{s}$$

$$I(s) = \frac{V/L}{s^2 + \frac{Rs}{L} + 1/Lc} \qquad (13.2)$$

To find the roots Equation (13.2):

Let, $s^2 + \frac{Rs}{L} + \frac{1}{Lc} = 0$; this is quadratic equation such as $ax^2 + bx + c = 0$ roots of

above equation $s_1 = \frac{-R}{L} + \sqrt{\left(\frac{R^2}{L} - \frac{4}{Lc}\right)}$ $\quad s_1 = \frac{-b \pm \sqrt{b^2 - 4ac}}{2a}$

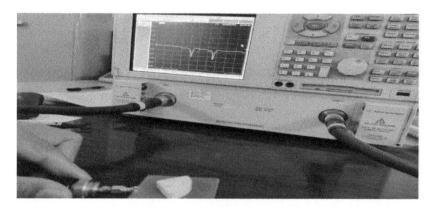

FIGURE 13.1   DRA equivalent circuit as R, L, C.

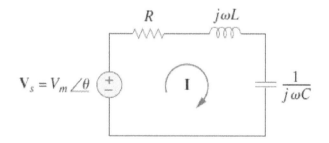

FIGURE 13.2   Series R, L, C circuit.

$$s_1 = \frac{-R}{2L} + \sqrt{\frac{R^2}{4L^2} - \frac{1}{Lc}}$$

Similarly, $s_2 = \dfrac{-R}{2L} - \sqrt{\dfrac{R^2}{4L^2} - \dfrac{1}{Lc}}$

1. Current $i(t)$ in Equation (13.2) can be written as:

$$I(s) = \frac{V/L}{(S - S_1)(S - S_2)} = \frac{A}{S - S_1} + \frac{A}{S - S_2} \qquad (13.3)$$

$$\frac{V}{L} = A(s - s_2) + B(s - s_1)$$

Case (1): Put $s = s_2$

$$B = \frac{V}{L(S_2 - S_1)}$$

Case (2): Put $s = s_1$

$$A = \frac{V}{L(S_1 - S_2)}$$

Now $s_2 - s_1 = -2\sqrt{\dfrac{R^2}{4L^2} - 1/LC}$

$$s_1 - s_2 = 2\sqrt{\frac{R^2}{4L^2} - 1/LC}$$

From Equation (13.3),

$$I(s) = \frac{V}{L}\left\{\frac{1}{(S_1 - S_2)(S - S_1)} + \frac{1}{(S_2 - S_1)(S - S_2)}\right\}$$

$$I(s) = 2\left\{\sqrt{\frac{R^2}{4L^2} - \frac{1}{Lc}}\right\}\frac{V}{L}\left\{\frac{1}{(S - S_1)} - \frac{1}{(S - S_2)}\right\}$$

Taking inverse Laplace transform,

$$i(t) = \frac{2V}{L}\sqrt{\frac{R^2}{4L^2} - \frac{1}{Lc}}\;\left\{e^{s_1 t} - e^{s_2 t}\right\}$$

2. Impedance

$$Z = R + X_L + X_c$$

$$Z = R + j\omega L - \frac{j}{\omega C}$$

$$Z = R + j\left(\omega L - \frac{1}{\omega C}\right)$$

$$|Z| = \sqrt{R^2 + \left(\omega L - \frac{1}{\omega C}\right)^2}$$

Resonant frequency
At resonance:

$$\omega L = \frac{1}{\omega C}$$

$$\omega^2 = \frac{1}{LC}$$

$$\omega_r = \frac{1}{\sqrt{LC}}$$

$$2\pi f_r = \frac{1}{\sqrt{LC}}$$

$$f_r = \frac{1}{2\pi\sqrt{LC}}$$

3. Bandwidth

R, L, C series circuit

$$f^2 = \frac{1}{2\pi L \times 2\pi c} = \frac{1}{4\pi^2 LC}$$

$$f_r = \frac{1}{2\pi\sqrt{LC}}$$

$$\omega_r = \frac{1}{\sqrt{LC}}$$

$$Q = \frac{X_L}{R}$$

$$Z = \sqrt{R^2 + \left(\omega L - \frac{1}{\omega C}\right)^2}$$

The current at resonance will be maximum

$$i_m = \frac{V_m}{2} ; \quad \tan\phi = \frac{X_C - X_L}{R} ; \quad \phi = \tan^{-1}\left(\frac{X_C - X_L}{R}\right)$$

4. Power

$$P = Vi = (V_m \sin\omega t)\,(i_m \sin(\omega t + \phi))$$

$$= \frac{V_m i_m}{2} \left[\cos\phi - \omega_s\left[2\omega t + \phi\right]\right]$$

$$\Rightarrow P = \frac{V_m i_m}{2}\cos\phi$$

$$= \frac{V_m}{\sqrt{2}}\,\frac{i_m}{\sqrt{2}}\,\cos\phi$$

$$P = VI\cos\phi \quad (V \text{ and } I \text{ are } RMS \text{ values})$$

Or

$$P = I^2 R\,\cos\phi$$

13.1.1 BANDWIDTH AT RESONANCE

Current at resonance is maximum.

$$i = \frac{V}{\sqrt{R^2 + \left(\omega L - \dfrac{1}{\omega C}\right)^2}} = \frac{i_{max}}{\sqrt{2}} = \frac{V}{R\sqrt{2}};$$

Hence, $\sqrt{R^2 + \left(\omega_1 L - \dfrac{1}{\omega_1 C}\right)^2} = R\sqrt{2}$

$$R^2 + \left(\omega_1 L - \frac{1}{\omega_1 C}\right)^2 = 2R^2$$

Let $\omega_1 = \omega$

$$\Rightarrow \left(\omega L - \frac{1}{\omega C}\right)^2 = R^2$$

$$\Rightarrow \left(\omega L - \frac{1}{\omega C}\right)^2 = \pm R$$

$$\Rightarrow LC\omega^2 - RC\omega - 1 = 0$$

$$\Rightarrow \omega = \frac{RC \pm \sqrt{R^2 C^2 + 4LC}}{2LC}$$

$$\omega_H = \frac{RC + \sqrt{R^2 C^2 + 4LC}}{2LC}$$

Similarly, taking $-R$ condition,

$$\omega_L = \frac{-RC + \sqrt{R^2 C^2 + 4LC}}{2LC}$$

So, BW $= \omega_H - \omega_L = \dfrac{RC}{2LC} + \dfrac{RC}{2LC}$

$$= \frac{2RC}{2LC} = \frac{R}{L}$$

So, bandwidth $= \dfrac{R}{L}$ Radian/sec (Figure 13.3)

$$BW = \frac{R}{2\pi L} Hz$$

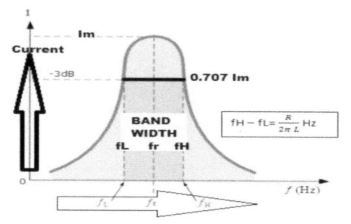

Series Resonance R,L,C circuit Band Width

FIGURE 13.3   Bandwidth of *R, L, C* circuit.

MATLAB PROGRAM FOR SLOUTION OF DRA AT RESONANCE

```
clc;
close all;
clear all;
w=-100:0.1:100;
R=input('enter the value of R=');
L=input('enter the value of L=');
C=input('enter the value of C=');
Vm=input('enter the value of vm=');
x1=w*L;
x2=1./(w*C);
x3=R;
z=sqrt(R^2+(((w*L)-(1\(w*C))).^2));
z1=Vm./z;
figure;
plot(x1)
hold on
plot(x2,'r')
hold on
plot(x3)
figure;
plot(z)
title('impedance vs frequency');
xlabel('w');
ylabel('|z|');
hold on;
figure;
plot(z1)
title('current vs frequency');
xlabel('w');ylabel('|I|');
```

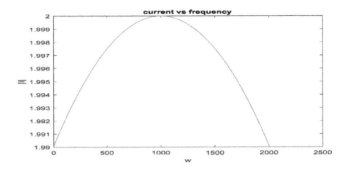

FIGURE 13.4   Bandwidth of *R, L, C* circuit.

INPUT IMPEDANCE
```
clc;
clear all;
close all;
r=input('Resistance(R) =');
l=input('Inductance(L)=');
c=input('Capacitance(C)=');
v=input('Voltage=');
f=5:75;
xl=2*pi*f*l;
xc=(1./(2*pi*f*c));
x=xl-xc;
z=sqrt((r^2)+(x.^2));
i=v./z;
vl=i.*(2*pi*f*l)
%plotting the graph%

% plot(f,r, '+');
%    hold on
%    plot(f,xl, 'b');
%    hold on
%    plot(f,xc, 'c');

subplot(2,2,1);
plot(f,xl);
grid;
xlabel('Frequency(Hz)');
ylabel('XL');
subplot(2,2,2);
plot(f,xc);
grid;
xlabel('Frequency(Hz)');
ylabel('XC');
subplot(2,2,3);
plot(f,z);
grid;
xlabel('Frequency(Hz)');
```

```
ylabel('Z');
subplot(2,2,4);
plot(f,i);
grid;
xlabel('frequency(Hz)');
ylabel('I(A)');
figure;
plot(vl)
```

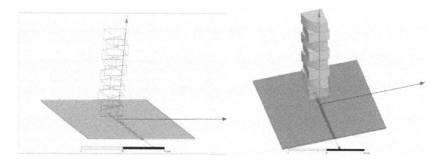

FIGURE 13.5   Stacked DRA HFSS model.

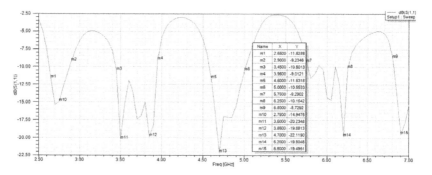

FIGURE 13.6   S$_{11}$ of stacked DRA showing multiple modes.

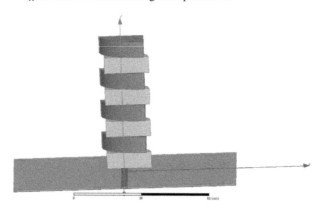

FIGURE 13.7   Eight stacked DRA showing multiple modes.

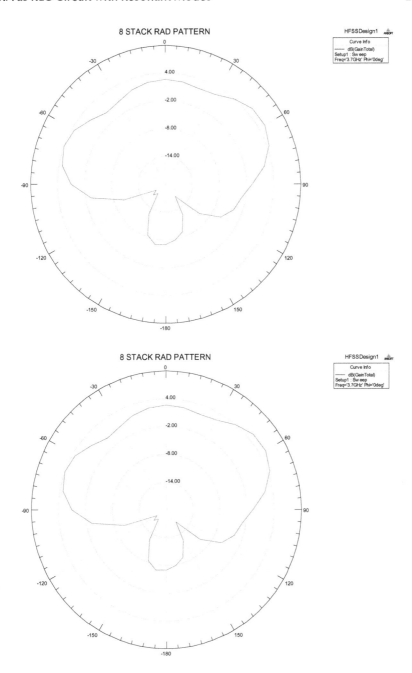

FIGURE 13.8 Radiation pattern of DRA.

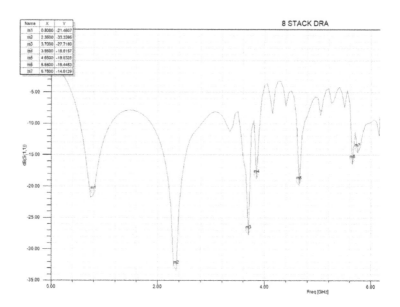

Name	X	Y
m1	0.8000	-21.4607
m2	2.3600	-33.3396
m3	3.7000	-27.7180
m4	3.8500	-18.6157
m5	4.6500	-19.8328
m6	5.6500	-18.4483
m7	5.7800	-14.6129

FIGURE 13.9 S_{11} of stacked DRA showing multiple modes.

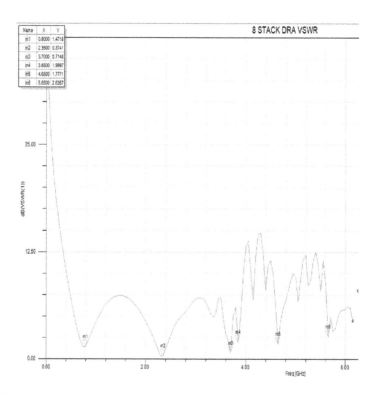

Name	X	Y
m1	0.8000	1.4718
m2	2.3600	0.3741
m3	3.7000	0.7148
m4	3.8500	1.9997
m5	4.6500	1.7771
m6	5.6500	2.0357

FIGURE 13.10 VSWR of DRA.

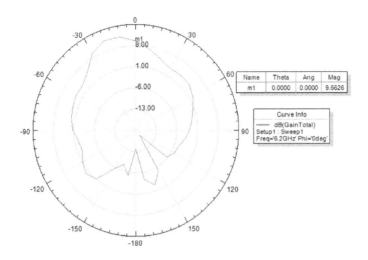

FIGURE 13.11 Radiation pattern of stacked DRA.

FIGURE 13.12 3D radiation pattern.

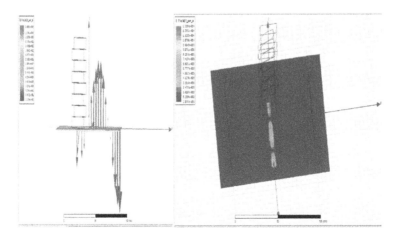

FIGURE 13.13 E-field vector.

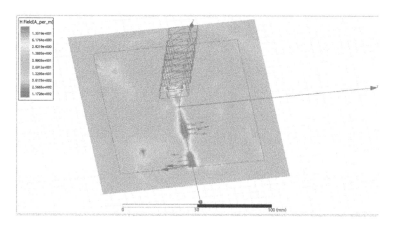

FIGURE 13.14 H-field vector of eight stack DRA.

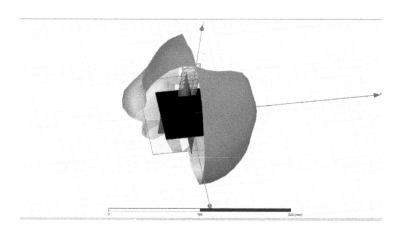

FIGURE 13.15 3D radiated fields.

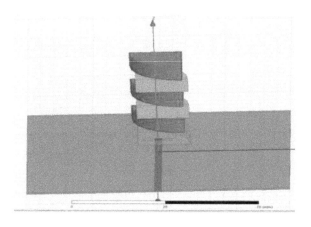

FIGURE 13.16 Six stack DRA.

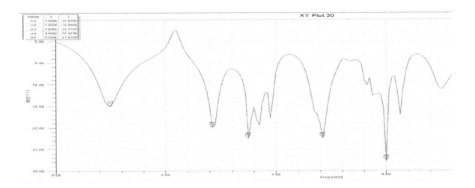

FIGURE 13.17 S_{11} of six stack DRA tuned.

13.2 DRA EQUIVALENT CIRCUIT WITH RESONANT MODES AND FIELD VECTORS

The DRA has resonant circuits with fundamental resonant frequencies f 01, second resonance at f 02 and other higher order resonance at f 03 and f 04, and so on. The parallel R, L, C circuit can be tuned to resonate at the desired frequency. The mutual inductance has coupling coefficient k. DRA is coupled to slot through microstrip line, which is used to feed DRA through slot from a source of 50 Ω impedance (Figure 13.18).

13.2.1 EQUIVALENT CCT

Input applied to DRA and its solution are shown in Figure 13.19.

$$V_i = V_m \sin \omega t$$

$$\frac{V_i}{V_s} = \frac{1}{n}$$

$$V_s = n \, V_i$$

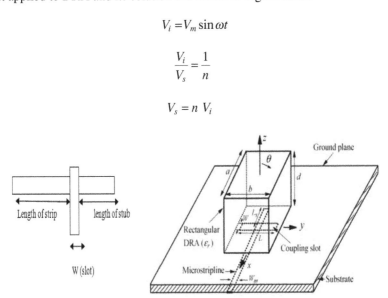

FIGURE 13.18 DRA with microstrip line, slot, stub.

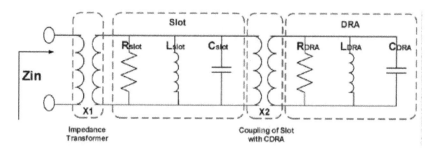

FIGURE 13.19   Stacked DRA at fundamental and higher order modes.

$$V = I Z_s = \frac{I}{Y_s}$$

$$V_o = n^2 V_{in}$$

$$\frac{I_{in}}{I_s} = \frac{n}{1} \Rightarrow I_s = \frac{I_{in}}{n}$$

$$\omega_s = \frac{1}{\sqrt{L_s C_s}}$$

$$I_s = \frac{V_m \sin \omega t}{R.n}$$

$$\text{Quality factor } (Q) = \frac{2\pi \times \text{Maximum energy stored}}{\text{Power disspated per cycle}}$$

$$Q = \frac{2\pi \left(\frac{1}{2} C_s V_s^2 + \frac{1}{2} Li (t)^2 \right)}{\left(\frac{V_s}{\sqrt{2}} \right)^2 \frac{1}{R_s} \times T}$$

$$= 2\pi \left(\frac{1}{2} n^2 C_s V_m^2 \sin^2 \omega_s t + \frac{1}{2} Ln^2 \frac{V_m}{\omega_s^2 L_s^2} \cos^2 \omega_s t \right)$$

$$I_s (t) = \frac{1}{2} \int n V_m \sin \omega t \, dt = \frac{n V_m}{L \omega_r} \cos \omega_r t$$

$$= 2\pi \left[\frac{1}{2} \, n^2 C_s V_m^2 \sin^2 \omega_s t + \frac{1}{2} \frac{n^2 V_m}{2 L_2 \omega_s^2} L \cos^2 \omega_s t \right]$$

$$\left(\frac{n V_m}{\sqrt{2}} \right)^2 \times \frac{1}{R} \times T$$

$$= 2\pi \left[\frac{1}{2} n^2 C_s V_m^2 \sin^2 \omega_s t + \frac{1}{2} \, n^2 C_s^2 V_m^2 \cos^2 \omega_s t \right]$$

$$Q = \frac{C_s \, 2\pi}{\dfrac{T}{R}} = C_s R \omega_s$$

$$Q = R_s \sqrt{\frac{C_s}{L_s}}$$

$$Z_T = Z_{d_1} + Z_c + Z_{d_2}$$

$$V_i = V_m \sin \omega t$$

$$Z_T = \frac{1}{Y_{d_1}} + Z_c + \frac{1}{Y_{d_2}}$$

$$V_o = I_o \, Z_T$$

$$Z_T = \frac{1}{\dfrac{1}{R_{d_1}} + j\left(\omega C_{d_1} - \dfrac{1}{\omega L_{d_1}}\right)} + j\left(\omega L_{d_1} - \dfrac{1}{\omega C_{d_1}}\right) + \frac{1}{\dfrac{1}{R_{d_2}} + j\left(\omega C_{d_2} - \dfrac{1}{\omega L_{d_2}}\right)}$$

$$Z_T = \frac{\dfrac{1}{R_{d_1^2}} - j\left(\omega C_{d_1} - \dfrac{1}{\omega L_{d_1}}\right)}{\dfrac{1}{R_{d_1^2}} + \left(\omega C_{d_1} - \dfrac{1}{\omega L_{d_1}}\right)^2} + J\left(\omega L_{d_1} - \dfrac{1}{\omega C_{d_1}}\right) + \frac{\dfrac{1}{R_{d_2^2}} - j\left(\omega C_{d_2} - \dfrac{1}{\omega L_{d_2}}\right)}{\dfrac{1}{R_{d_2^2}} + \left(\omega C_{d_2} - \dfrac{1}{\omega L_{d_2}}\right)^2}$$

$$I_{ma}\left[Z_T\right] = J\left[\left(\omega L_s - \dfrac{1}{\omega C_s}\right) \frac{-\left(\omega C_{d_1} - \dfrac{1}{\omega L_{d_1}}\right)}{\dfrac{1}{R_{d_1^2}} + \left(\omega C_{d_1} - \dfrac{1}{\omega L_{d_1}}\right)^2} \frac{\left(\omega C_{d_2} - \dfrac{1}{\omega L_{d_2}}\right)}{\dfrac{1}{R_{d_2^2}} + \left(\omega C_{d_2} - \dfrac{1}{\omega L_{d_2}}\right)^2}\right]$$

$$\text{Real}\left[Z_T\right] = \frac{\dfrac{1}{R_{d_1^2}}}{\dfrac{1}{R_{d_1^2}} + \left(\omega C_{d_1} - \dfrac{1}{\omega L_{d_1}}\right)^2} + \frac{\dfrac{1}{R_{d_2^2}}}{\dfrac{1}{R_{d_2^2}} + \left(\omega C_{d_2} - \dfrac{1}{\omega L_{d_2}}\right)^2}$$

For resonance frequency (Imaginary port $\Rightarrow \times = 0$)

$$\omega L_s - \frac{1}{\omega C_s} = 0 \Rightarrow \omega_c = \frac{1}{\sqrt{L_s \, C_s}}$$

$$\omega C_{d_1} - \frac{1}{\omega L_{d_1}} = 0 \Rightarrow \omega_{d_1} = \frac{1}{\sqrt{L_{d_1} \, L_{d_1}}}$$

$$\omega C_{d_2} - \frac{1}{\omega L_{d_2}} = 0 \Rightarrow \omega_{d_2} = \frac{1}{\sqrt{L_{d_2} \, L_{d_2}}}$$

$$\text{Real}(Z_T) = \frac{1}{1 + \left(\omega C_{d_1} - \dfrac{1}{\omega L_{d_1}}\right)^2 R_{d_1^2}} + \frac{1}{1 + \left(\omega C_{d_2} - \dfrac{1}{\omega L_{d_2}}\right)^2 R_{d_2^2}}$$

$$\text{Dynamic impedance}(Z_T) = \frac{1}{1 + \left(\omega C_{d_1} - \dfrac{1}{\omega L_{d_1}}\right)^2 R_{d_1^2}} + \frac{1}{1 + \left(\omega C_{d_2} - \dfrac{1}{\omega L_{d_2}}\right)^2 R_{d_1^2}}$$

$$\text{Imaginary}(Z_T) = \left[\left(\omega_{L_s} - \frac{1}{\omega_{L_s}}\right) - \frac{\left(\omega C_{d_1} - \dfrac{1}{\omega L_{d_1}}\right)}{\dfrac{1}{R_{d_1^2}} + \left(\omega C_{d_1} - \dfrac{1}{\omega L_{d_1}}\right)^2} - \frac{\left(\omega C_{d_2} - \dfrac{1}{\omega L_{d_2}}\right)}{\dfrac{1}{R_{d_2^2}} + \left(\omega C_{d_2} - \dfrac{1}{\omega L_{d_2}}\right)^2} \right]$$

In equivalent circuit, voltage is constant at all times, the effect due to transformer coupling factor (N_1/N_2) is dominant.

$$V_o = I_T \, Z_T$$

$$V_o = \frac{I_T}{Y_T}$$

$$Z_T = \frac{1}{Y_T}$$

Z_T has two components real and imaginary.

13.2.2 Special Case of DRA Equivalent Circuit

Solution of equivalent circuit: Here $R1$, $L1$, $C1$ are slot components now treated as R_S, L_S, C_S, and $R2$, $L2$, $C2$ are DRA components treated as $R_D L_D C_D$.

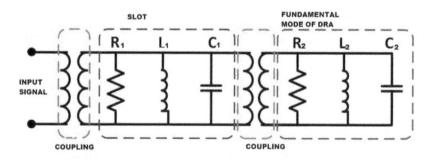

EQUIVALENT CIRCUIT OF DRA

Case 1:

$$Y_D = \frac{1}{R_D} + \frac{1}{s L_D} + s C_D$$

$$Y_s = \frac{1}{R_s} + \frac{1}{s L_s} + s C_s$$

$$Z_{inD} = \frac{1}{n^2} Z_D = \frac{1}{n^2} \frac{1}{Y_D} = \frac{1}{n^2 \left(\frac{1}{R_s} + \frac{1}{s L_s} + s C_s \right)}$$

$$Z_{inD} = \frac{s L_s R_s}{n^2 \left(R_s + s L_s + s^2 L_s C_s R_s \right)}$$

$$Z_s = \frac{1}{Y_s} + Z_{inD}$$

$$Z_s = \frac{1}{\frac{1}{R_s} + \frac{1}{s L_s} + s C_s} + Z_{in}$$

$$Z_s = \frac{R_s \, s \, L_s}{s^2 L_s C_s R_s + R_s + s \, L_s} + \frac{s L_D R_D}{n^2 \left(R_D + s L_D + s^2 L_D C_D R_D \right)}$$

$$Z_{inD} = \left\{ R_s L_s s \left\{ n^2 \left(s^2 L_D R_D C_D + s L_D + R_D \right) \right\} + s L_D R_D \left(s^2 L_s R_s C_s s L_s R_s \right) \right\} / $$
$$\left\{ n^2 \left(s^2 L_D R_D C_D + s L_D + R_D \right) \left(s^2 L_s R_s C_s + s L_s + R_s \right) \right\}$$

$$Z_{in} = \frac{1}{n^2} Z_{inD}$$

$$Z_{in} = \frac{1}{n^4} \left[n^2 \, s^3 L_D L_s R_s \, R_D \, C_D + n^2 \, s^2 L_D L_s R_s + s R_s \, R_D \, L_s \right.$$
$$\left. + s^3 L_D R_D L_s R_s C_s + s^2 L_D R_D L_s + s L_D R_D \right] / $$
$$\begin{bmatrix} (s^4 L_D \, R_D \, C_D L_s R_s \, C_s + s^2 \left(L_D \, R_D \, C_D R_D + L_s R_s \, C_s R_s \right) \\ + S^3 \left(L_D R_D C_D L_s + L_D L_s R_s C_s \right) + R_s R_D + s \left(L_D R_s + L_s R_D \right) \end{bmatrix}$$

$$Z_{in} = \frac{1}{n^4} \left[\left(s^3 \left(L_D R_s L_s R_D C_D \, A^2 + L_D R_D \, L_s R_s \, C_s \right) + s^2 \left(n^2 L_D L_s \, R_s + L_D R_D \, L_s \right) \right. \right.$$
$$\left. + s \left(L_D R_D \, R_s + R_s R_D L_s \right) \right] / $$
$$\left[(s^4 L_D \, R_D \, C_D L_s R_s \, C_s) + s^3 \left(L_D R_D C_D L_s + L_D L_s R_s \, C_s \right) + s^2 \right] $$
$$\left(L_D R_D C_D R_D + L_s R_s \, C_s R_s \right) + S \left(L_D R_s + L_s R_D \right) + R_s R_D \right]$$

Case 2: Higher order modes in DRA:

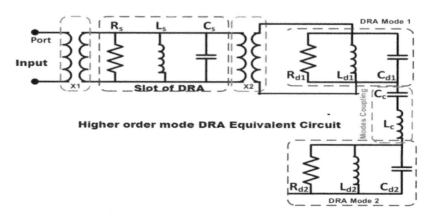

Higher order mode DRA Equivalent Circuit

$$V_i(t) = V_m \sin \omega t$$

$$\frac{V_i}{V_s} = \frac{1}{n}$$

$$V_s = n\, V_i$$

$$V = I Z_s = \frac{I}{Y_s}$$

$$V_o = n^2 V_{in}$$

$$\frac{I_{in}}{I_s} = \frac{n}{1} \Rightarrow I_s = \frac{I_{in}}{n}$$

$$\omega_s = \frac{1}{\sqrt{L_s C_s}}$$

$$I_s = \frac{V_m \sin \omega t}{R.n}$$

$$\text{Quality factor } (Q) = \frac{2\pi \times \text{Maximum energy stored}}{\text{Power disspated per cycle}}$$

$$Q = \frac{2\pi \left(\frac{1}{2} C_s V_s^2 + \frac{1}{2} Li\,(t)^2 \right)}{\left(\frac{V_s}{\sqrt{2}} \right)^2 \frac{1}{R_s} \times T}$$

$$= 2\pi \left(\frac{1}{2} n^2 C_s V_m^2 \sin^2 \omega_s t + \frac{1}{2} \mathrm{Ln}^2 \frac{V_m}{\omega_s^2 L_s^2} \cos^2 \omega_s t \right)$$

$$I_s(t) = \frac{1}{L} \int n V_m \sin \omega t \, dt = \frac{n V_m}{L \omega_r} \cos \omega_r t$$

$$Q = 2\pi \left[\frac{1}{2} \ n^2 C_s V_m^2 \sin^2 \omega_s t + \frac{1}{2} \frac{n^2 V_m}{L_2 \omega_s^2} L \cos^2 \omega_s t \right] / \left(\frac{n V_m}{\sqrt{2}} \right)^2 \times \frac{1}{R} \times T$$

$$= 2\pi \left[\frac{1}{2} n^2 C_s V_m^2 \sin^2 \omega_s t + \frac{1}{2} \ n^2 C_s^2 V_m^2 \cos^2 \omega_s t \right] / \left(\frac{n V_m}{\sqrt{2}} \right)^2 \times \frac{1}{R} \times T$$

$$Q = \frac{C_s \, 2\pi}{\dfrac{T}{R}} = C_s R \omega_s$$

$$Q = R_s \sqrt{\frac{C_s}{L_s}}$$

$$Z_T = Z_{d_1} + Z_c + Z_{d_2}$$

$$V_i = V_m \sin \omega t$$

$$Z_T = \frac{1}{Y_{d_1}} + Z_c + \frac{1}{Y_{d_2}}$$

$$V_o = I_o Z_T$$

$$Z_T = \frac{1}{\dfrac{1}{R_{d_1}} + j\left(\omega C_{d_1} - \dfrac{1}{\omega L_{d_1}} \right)} + j\left(\omega L_{d_1} - \frac{1}{\omega C_{d_1}} \right) + \frac{1}{\dfrac{1}{R_{d_2}} + j\left(\omega C_{d_2} - \dfrac{1}{\omega L_{d_2}} \right)}$$

$$Z_T = \frac{\dfrac{1}{R_{d_1^2}} - j\left(\omega C_{d_1} - \dfrac{1}{\omega L_{d_1}} \right)}{\dfrac{1}{R_{d_1^2}} + \left(\omega C_{d_1} - \dfrac{1}{\omega L_{d_1}} \right)^2} + J\left(\omega L_{d_1} - \frac{1}{\omega C_{d_1}} \right) + \frac{\dfrac{1}{R_{d_2^2}} - j\left(\omega C_{d_2} - \dfrac{1}{\omega L_{d_2}} \right)}{\dfrac{1}{R_{d_2^2}} + \left(\omega C_{d_2} - \dfrac{1}{\omega L_{d_2}} \right)^2}$$

$$I_{m_a}\left[Z_T \right] = J \left[\left(\omega L_c - \frac{1}{\omega C_c} \right) \frac{-\left(\omega C_{d_1} - \dfrac{1}{\omega L_{d_1}} \right)}{\dfrac{1}{R_{d_1^2}} + \left(\omega C_{d_1} - \dfrac{1}{\omega L_{d_1}} \right)^2} \quad \frac{\left(\omega C_{d_2} - \dfrac{1}{\omega L_{d_2}} \right)}{\dfrac{1}{R_{d_2^2}} + \left(\omega C_{d_2} - \dfrac{1}{\omega L_{d_2}} \right)^2} \right]$$

$$\text{Real}[Z_T) = \frac{\dfrac{1}{R_{d_1^2}}}{\dfrac{1}{R_{d_1^2}} + \left(\omega C_{d_1} - \dfrac{1}{\omega L_{d_1}}\right)^2} + \frac{\dfrac{1}{R_{d_2^2}}}{\dfrac{1}{R_{d_2^2}} + \left(\omega C_{d_2} - \dfrac{1}{\omega L_{d_2}}\right)^2}$$

For resonance frequency (Imaginary part $\Rightarrow \times = 0$)

$$\omega L_c - \frac{1}{\omega C_c} = 0 \Rightarrow \omega_c = \frac{1}{\sqrt{L_c\, C_c}}$$

$$\omega C_{d_1} - \frac{1}{\omega L_{d_1}} = 0 \Rightarrow \omega_{d_1} = \frac{1}{\sqrt{L_{d_1}\, L_{d_1}}}$$

$$\omega C_{d_2} - \frac{1}{\omega L_{d_2}} = 0 \Rightarrow \omega_{d_2} = \frac{1}{\sqrt{L_{d_2}\, L_{d_2}}}$$

$$\text{Real}(Z_T) = \frac{1}{1 + \left(\omega C_{d_1} - \dfrac{1}{\omega L_{d_1}}\right)^2 R_{d_1^2}} + \frac{1}{1 + \left(\omega C_{d_2} - \dfrac{1}{\omega L_{d_2}}\right)^2 R_{d_2^2}}$$

$$\text{Dynamic impedance } (Z_T) = \frac{1}{1 + \left(\omega C_{d_1} - \dfrac{1}{\omega L_{d_1}}\right)^2 R_{d_1^2}} + \frac{1}{1 + \left(\omega C_{d_2} - \dfrac{1}{\omega L_{d_2}}\right)^2 R_{d_1^2}}$$

$$\text{Imaginary } (Z_T) = \left[\left(\omega_{L_c} - \frac{1}{\omega_{L_c}}\right) - \frac{\left(\omega C_{d_1} - \dfrac{1}{\omega L_{d_1}}\right)}{R_{d_1^2} + \left(\omega C_{d_1} - \dfrac{1}{\omega L_{d_1}}\right)^2} - \frac{\left(\omega C_{d_2} - \dfrac{1}{\omega L_{d_2}}\right)}{\dfrac{1}{R_{d_2^2}} + \left(\omega C_{d_2} - \dfrac{1}{\omega L_{d_2}}\right)^2}\right]$$

In equivalent circuit, voltage is constant at all times, transformer coupling factor (N_1/N_2) is dependent on many parameters.

$$V_o = I_T\, Z_T$$

$$V_o = \frac{I_T}{Y_T}$$

$$Z_T = \frac{1}{Y_T}$$

Z_T has two components:

1. Real components
2. Imaginary component

Case 3:

$$\text{Quality factor } (Q) = R_s \sqrt{\frac{C_s}{L_s}}$$

and

$$\text{Bandwidth} = \frac{1}{R_s C_s}$$

And resonance frequency $(\omega_s) = \dfrac{1}{\sqrt{L_s C_s}}$

Calculation for bandwidth for the first circuit is,

$$V_s = nV_i = nV_m \sin \omega t$$

$$V_s(t) = \frac{i_s(t)}{Y_s} = \frac{i_s(t)}{\dfrac{1}{R_s} + J\left(\omega C_s - \dfrac{1}{\omega L_s}\right)}$$

where $i_s\, t = \dfrac{I_{in}}{n}$

$$V_s(t) = \frac{I_{in}}{n}\left(\frac{1}{\dfrac{1}{R_s} + J\left(\omega C_s - \dfrac{1}{\omega L_s}\right)}\right)$$

$$V_s(t) = \frac{I_m \sin \omega t}{n\left[\dfrac{1}{R_s} + J\left(\omega C_s - \dfrac{1}{\omega L_s}\right)\right]}$$

$$V_s(t) = \frac{\dfrac{I_m}{n} < 0}{\sqrt{\dfrac{1}{Rs^2} + \left(\omega C_s - \dfrac{1}{\omega L_s}\right)^2}\left(< \tan^{-1}\left(\dfrac{\omega C_s - \dfrac{1}{\omega L_s}}{G_s}\right)\right)}$$

$$V_s(t) = \frac{\dfrac{I_m}{n}}{\sqrt{\dfrac{1}{Rs^2} + \left(\omega C_s - \dfrac{1}{\omega L_s}\right)^2}} < \tan^{-1}\left(\frac{\omega C_s - \dfrac{1}{\omega L_s}}{\dfrac{1}{Rs}}\right)$$

$$|V_s(\omega)| = \frac{I_m / n}{\sqrt{\dfrac{1}{Rs^2} + \left(\omega C_s - \dfrac{1}{\omega L_s}\right)^2}}$$

$A + \omega C_s - \dfrac{1}{\omega L_s} = 0$; Then $|V_s|_{\max}$

For $\dfrac{V_s|_{\max}}{\sqrt{2}}$ two frequencies are present: lower cut-off and upper cut-off frequency.

Frequency between cut-off is called bandwidth.

ω_1 is called a lower cut-off frequency and ω_2 is called a higher cut-off frequency.

$$\frac{I_m}{n\sqrt{\dfrac{1}{Rs^2} + \left(\omega C_s - \dfrac{1}{\omega L_s}\right)2}} = \frac{I_m}{n\sqrt{\dfrac{1}{Rs^2} + (\pm\dfrac{1}{Rs})^2}}$$

$$\omega C_s - \frac{1}{\omega L_s} = -\frac{1}{R_s} \tag{13.4}$$

$$\omega C_s + \frac{1}{\omega L_s} = \frac{1}{R_s} \tag{13.5}$$

For lower cut-off frequency, let us put $\omega = \omega_1$ in Equation (13.4).

$$\omega_1 C_s - \frac{1}{\omega_1 L_s} = -\frac{1}{R_s}$$

$$\omega_1^2 L_s C_s - 1 = -\omega_1 L_s G_s$$

$$\omega_1^2 L_s C_s + \omega_1 L_s G_s - = 0$$

$$\omega_1 = \frac{-L_s G_s \pm \sqrt{(L_s G_s)^2 + 4 L_s C_s}}{2 L_s C_s}$$

$$\omega_1 = \frac{-G_s}{2C_s} \pm \sqrt{\left(\frac{G_s}{2C_s}\right)^2 + \frac{1}{L_s C_s}}$$

Similarly, for higher cut-off frequency, we put $\omega = \omega_2$ for Equation (13.5), then

$$\omega_1^2 L_s C_s - \omega_1 L_s R_s - 1 = 0$$

$$\omega_2 = \frac{-G_s}{2C_s} \pm \sqrt{\left(\frac{G_s}{2C_s}\right)^2 + \frac{1}{L_s C_s}}$$

Summarize equations (13.4) and (13.5),

$$\omega_1 = -\frac{G_s}{2C_s} + \sqrt{\left(\frac{G_s}{2C_s}\right)^2 + \frac{1}{L_s C_s}}$$

$$\omega_2 = \frac{G_s}{2C_s} + \sqrt{\left(\frac{G_s}{2C_s}\right)^2 + \frac{1}{L_s C_s}}$$

For bandwidth (BW) $= \omega_2 - \omega_1$

$$= \frac{G_s}{2C_s} + \sqrt{\left(\frac{G_s}{2C_s}\right)^2 + \frac{1}{L_s C_s}} + \frac{G_s}{2C_s} - \sqrt{\left(\frac{G_s}{2C_s}\right)^2 + \frac{1}{L_s C_s}}$$

$$\text{Bandwidth} = \frac{2G_s}{2C_s} = \frac{G_s}{C_s} = \frac{1}{R_s C_s};$$

$$\omega_r = \frac{1}{L_s C_s}$$

$$\text{Bandwidth} = \frac{G_s}{C_s} = \frac{1}{R_s C_s} = \frac{1}{C_s}$$

$$\text{Quality factor}(Q_s) = R_s \sqrt{\frac{C_s}{L_s}}$$

$$Q_s = R_s W_s C_s$$

$$Q_s = \frac{W_s}{B.w}; \quad \text{Quality factor} = \frac{\text{Resonance frequency}}{\text{Bandwidth}}$$

$$Q = \frac{f_r}{f_2 - f_1}$$

Case 4:
For a simplified circuit, the solution will become:

$$\frac{V_{in}}{V_s} = \frac{1}{n} \quad\quad V_s = n V_{in}$$

$$\frac{V_s}{V_o} = \frac{1}{n} \quad\quad V_o = n V_s$$

$$\Rightarrow V_o = n^2 V_{in}$$

But using the CKT $V_o = I_o Z_T$

$$I_o \times Z_T \; n^2 V_{in}$$

$$I_o = \frac{n^2 V_m \sin \omega t}{Z_T}$$

where $Z_T = Z_{d_1} + Z_c + Z_{d_2}$

where $Z_{d_1} \Rightarrow DRA - 1$ total impedance

$$Z_c \Rightarrow \text{Coupling impedance}$$

$$Z_{d_2} \Rightarrow DRA - 2 \;\; \text{Total impedance}$$

$$Z_T = \frac{1}{Y_{d_1}} + Z_c + \frac{1}{Y_{d_2}}$$

$$Z_T = \frac{1}{\dfrac{1}{R_{d_1}} + J\left(\omega c_{d_1} - \dfrac{1}{\omega L_{d_1}}\right)} + J\left(\omega L_c - \frac{1}{\omega C_c}\right) + \frac{1}{\dfrac{1}{R_{d_2}} + J\left(\omega c_{d_2} - \dfrac{1}{\omega L_{d_2}}\right)}$$

$$I_o = n^2 V_m \omega t \cdot$$

$$\frac{\dfrac{1}{R_{d_1}} - j\left(\omega c_{d_1} - \dfrac{1}{\omega L_{d_1}}\right)}{\dfrac{1}{R_{d_1^2}} + \left(\omega c_{d_1} - \dfrac{1}{\omega L_{d_1}}\right)^2} + J\left(\omega L_c - \frac{1}{\omega C_c}\right) + \frac{\dfrac{1}{R_{d_2}} - J\left(\omega c_{d_2} - \dfrac{1}{\omega L_{d_2}}\right)}{\dfrac{1}{R_{d_2^2}} + \left(\omega c_{d_2} - \dfrac{1}{\omega L_{d_2}}\right)^2}$$

collecting real and imaginary port separately.

$$I_o = \frac{n^2 V_m \omega t}{\left[\dfrac{\dfrac{1}{R_{d_1}}}{P} + \dfrac{\dfrac{1}{R_{d_2}}}{Q} + J\left(\left(\omega L_c - \dfrac{1}{\omega C_c}\right) - \dfrac{\left(\omega c_{d_1} - \dfrac{1}{\omega L_{d_1}}\right)}{P} - \dfrac{\left(\omega c_{d_2} - \dfrac{1}{\omega L_{d_2}}\right)}{Q}\right)\right.}$$

where, $P = \dfrac{1}{R_{d_1^2}} + \left(\omega c_{d_1} - \dfrac{1}{\omega L_{d_1}}\right)^2$

$$Q = \frac{1}{R_{d_2^2}} + \left(\omega c_{d_2} - \frac{1}{\omega L_{d_2}}\right)^2$$

for the resonance frequency for different modes.
When imaginary part = 0; at resonance

$$\left(\omega L_c - \frac{1}{\omega C_c}\right) - \frac{\left(\omega C_{d_1} - \frac{1}{\omega L_{d_1}}\right)}{P}\left(\omega C_{d_2} - \frac{1}{\omega L_{d_2}}\right)}{Q}$$

For the $\omega_{d_1} = \dfrac{1}{\sqrt{L_{d_1} C_{d_1}}}$

$$\omega_c = \frac{1}{\sqrt{L_c C_c}}$$

$$\omega_{d_2} = \frac{1}{\sqrt{L_{d_2} C_{d_2}}}$$

At different mode of resonance, the dynamic impedance or real port of impedance is

$$Z_D = Z_R = \frac{\frac{1}{R_{d_1}}}{Q} + \frac{\frac{1}{R_{d_2}}}{Q} = \frac{\frac{1}{R_{d_1}}}{\frac{1}{Rd_2^2} + \left(\omega C_{d_2} - \frac{1}{\omega L_{d_2}}\right)^2}$$

At the resonance for all modes, the output current (I_o) is maximum

$$I_o = (I_o)_{max} = \frac{n^2 V_m \sin\omega t}{\frac{1}{R_{d_1}} + \left[\frac{1}{R_{d_1}^2}\left(\omega C_{d_1} - \frac{1}{\omega L_{d_1}}\right)^2\right] \frac{1}{R_{d_2}} + \left[\frac{1}{R_{d_2}^2}\left(\omega C_{d_2} - \frac{1}{\omega L_{d_2}}\right)^2\right]}$$

$$(I_o)_{max} = \frac{n^2 V_m \sin\omega t}{\left[\frac{1}{R_{d_1}} + R_{d_1}\left(\omega C_{d_1} - \frac{1}{\omega L_{d_1}}\right)^2\right] \frac{1}{R_{d_2}} + \left[\frac{1}{R_{d_2}} + R_{d_2}\left(\omega C_{d_2} - \frac{1}{\omega L_{d_2}}\right)^2\right]}$$

due to combining all impedance the plot $Z_T V_s$ frequency (ω).

Then combing slot impedance, $Z_{in} = Z_s +$. Then transfer coupling factor transfer to Z_T

$$Z_{in} = \frac{1}{Y_s} + \frac{1}{\frac{1}{n^2}Z^T}$$

$$Z_{in} = Z_s + \frac{n^2}{Z^T}$$

$$Z_{in} = Z_s + n^2 Y_T$$

Offer this combing graph impedance V_s frequency (Radian/sec) impedance.

Offers combing the all mode the resistance capacitors and inductor value are changed with respect to the adding circuit.

In this case:

Resistance value decreases due to the parallel combination

Capacitance value increases due to the parallel combination

Total Inductance in parallel circuit decreases.

Quality factor of the CKT increases DRA are adding to different.

$$Q_s = R_s \sqrt{\frac{C_s}{L_s}} \text{ For slot time.}$$

$$Q_{d_1} = R_{d_1} \sqrt{\frac{C_{d_1}}{L_{d_1}}} \text{ For DRA-1}$$

$$Q_{d_2} = R_{d_2} \sqrt{\frac{C_{d_2}}{L_{d_2}}} \text{ For DRA-2}$$

$$C_{d_2} > C_{d_1} > C_s$$

and $L_{d_2} < L_{d_1} < L_s$

Quality factor increases and bandwidth of this CKT decreases:

$$\text{Bandwidth } (\text{BW}) = \frac{1}{R_s C_s} = \frac{1}{C_s} \text{ for slot}$$

$$\text{Bandwidth } (\text{BW}) = \frac{1}{R_{d_1} C_{d_1}} = \frac{1}{C_{d_1}} \text{ for DRA-1}$$

$$\text{Bandwidth } (\text{BW}) \frac{1}{R_{d_2} C_{d_2}} = \frac{1}{C_{d_2}} \text{ for DRA-2}$$

Time constant of CKT is $t_{d_2} > t_{d_1} > t_s$ because the capacitor increases.

$$\text{Quality factor} = \frac{\text{Resonance frequency}}{\text{Band width}}$$

Bandwidth is decreased due to increase in the DRA because the time constant is increased $(t) = \text{Rc}$.

Due to bandwidth decreases, the quality factor is increased and the selectivity of the signal is increased because selectivity is a proportional quality factor inversely to bandwidth. Higher order modes will result in higher currents in DRA. The Q factor is proportional to permittivity of material. Low Q will dissipate and high Q will store energy.

13.3 FIELDS PATTERN (FIGURES 13.20–13.22)

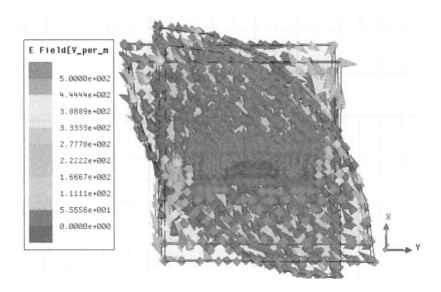

FIGURE 13.20 E-field DRA at 3.5 GHz.

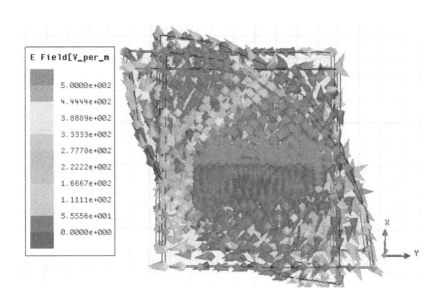

FIGURE 13.21 E-field at 4.7 GHz.

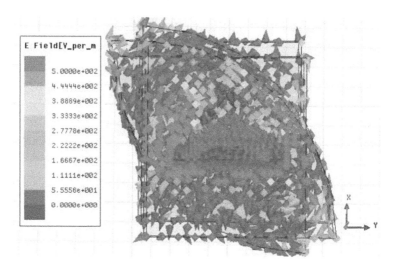

FIGURE 13.22   E-field at 6.2 GHz.

13.4   RESONANT MODES EXCITED IN DRA

Mode theory can provide insight about current distribution inside the DRA through orthogonal basis function of currents by making use of eigen vector theory. At any instant of time, total distributed currents in DRA are a weighted sum of eigen currents or superposition of excited modes in DRA. Surface current densities inside DRA are created by input excitations. The propagation of RF fields is based on Maxwell's curl equations. DRA seems to be a collection of R, L, C parallel tuned circuits arranged in a pattern such as super position of four or three vector valued basis functions.

$$\underline{E}(x,y,z,t) = \sum_{m,n,p=1}^{\infty} \mathrm{Re}\left\{c(mnp)e^{j\omega(mnp)t}\underline{\psi}_{mnp}^{E}(x,y,z)\right\}$$

$$+ \sum_{m,n,p=1}^{\infty} \mathrm{Re}\left\{d(mnp)e^{j\omega(mnp)t}\underline{\bar{\phi}}_{mnp}^{E}(x,y,z)\right\}$$

$$\underline{H}(x,y,z,t) = \sum_{m,n,p=1}^{\infty} \mathrm{Re}\left\{c(mnp)e^{j\omega(mnp)t}\underline{\psi}_{mnp}^{H}(x,y,z)\right\}$$

$$+ \sum_{m,n,p=1}^{\infty} \mathrm{Re}\left\{d(mnp)e^{j\omega(mnp)t}\underline{\bar{\phi}}_{mnp}^{H}(x,y,z)\right\}$$

where H_z and E_z are fields and D_{mnp} and C_{mnp} are coefficient magnitudes with linear combinations and their phase with indices at some known frequency.

This is shown in Figures 13.23 through 13.28.

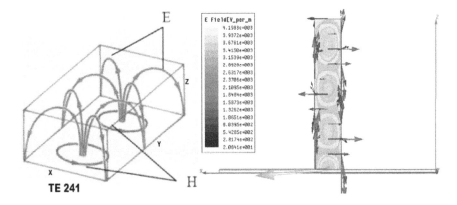

FIGURE 13.23 E_z and H_z as propagating fields.

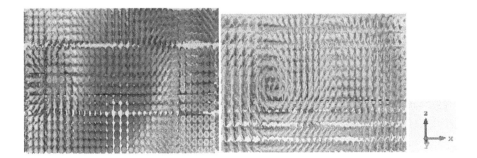

FIGURE 13.24 Modes in xz plane (TE_{211}) E and H vector field.

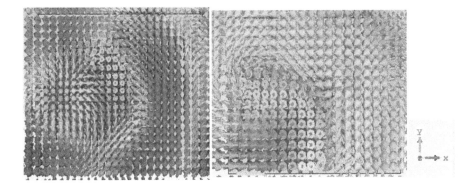

FIGURE 13.25 xy plane (TE_{211}) E and H vector field.

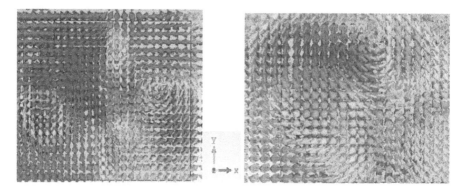

FIGURE 13.26   Resonant modes in xy plane (TE$_{121}$) E and H vector field.

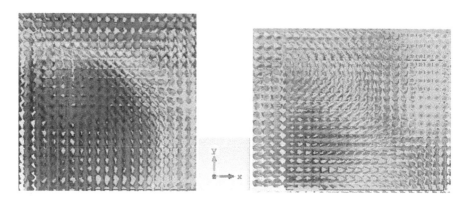

FIGURE 13.27   Resonant modes excited in yz plane (TE$_{121}$) E and H field.

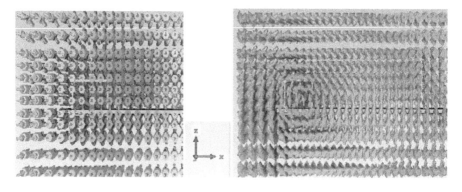

FIGURE 13.28   Fundamental mode field in zx plane (TE$_{111}$) E and H field.

13.5 CONCLUSION

Getiing multiple bands in single DRA is very useful for vehicular antenna designs. AM/FM with meanderline GPS/GSM/SDAR/SRR/DRA with shark fin type with tuner features can be embedded in single DRA.

REFERENCES

1. Glen Dash. How RF Anechoic Chambers Work. Copyright 1999, 2005 Ampyx LLC.
2. V. Rodriguez, Basic Rules for Anechoic Chamber Design, Part One: RF Absorber Approximations Ml Technologies, Suwanee, Ga. January 14, 2016.
3. Ferrite Tile Absorbers. Fair-rite Products Corporation.
4. C. A. Balanis, *Antenna Theory, Analysis and Design*. 3rd ed. Wiley, 2009.
5. L. H. Hemming, *Electromagnetic Anechoic Chambers: A Fundamental Design and Specifications Guide*, Piscataway, NJ: IEEE Press/New York: Wiley, 2002.
6. S. K. Das, A. Das. *Antenna and Wave Propogation*. New York, McGraw Hill, 2017.
7. J. Kraus, *Antennas*. 2nd ed. New York, McGraw-Hill, 1973.
8. H. King, F. Shimabukuro, J. Wong, *Characteristics of a Tapered Anechoic Chamber*. Los Angeles: Aerospace Corp., 1967.
9. B. T. DeWitt and W. D. Burnside, "Electromagnetic Scattering by Pyramidal and Wedge Absorber," *IEEE Trans. Antennas Propag.*, vol. 36, no. 7, pp. 971–984, 1988.

14 Multi Mode and Multiband DRA for Vehicular Applications

14.1 INTRODUCTION

Vehicular systems are a lifeline for people in urban areas. Present era vehicles have embedded automation features with the help of wireless RF sensors and biomedical sensors for infotainment and entertainment [1]. High-end cars have AM radio, FM radio, satellite radio, TPMS (tire pressure monitor), remote key entry, remote start, in-vehicle TV, DAB (digital audio broadcast), GPS (Global Positioning System), Bluetooth, collision avoidance radar, parking assist radar, electronic toll collection, Wi-Fi, Wi-Max, WLAN, Bluetooth, GSM (global system for mobile communication), car-to-car and car to infrastructure communication in smart cities, temperature sensor, traffic monitor, route map, and other wireless sensing capabilities for onboard patient biomedical monitoring [2]. All these wireless services need advanced wireless technologies. The wireless services become possible with the use of advanced antennas. This paper proposes an advanced dielectric resonator antenna for vehicular communications. This can be used to achieve car infotainment and entertainment services. HFSS simulation has been used for validating the concept and results of radiation, and scattering parameters have been obtained [3]. This DRA is suitable for Wi-Fi 3.6 and 5 GHz used in vehicular communication. Also, it can be useful for radar and satellites operating in the X-band frequency spectrum [4]. DRA at 24 GHz for LRR and SRR has also been proposed for vehicular networks. These mm wave antenna can also be used for 5G communication [5]. One challenge posed for vehicular antennas is directionality, or the ability to pick up the right level of signal with the vehicle facing different directions. The optimal solution for this problem is to use a diverse pattern of antennas, maybe multiple antennas or multiband antennas [6]. Moreover, if active antennas are used with the circuitry to accept the strongest desired signal, then they can be used for communication. RF systems are used for in-car communication with infrastructure [7]. Aerodynamics, safety and aesthetic design of the car are a few important parameters to keep in mind while designing these RF systems and car antennas [8]. The system gets choked when user density increases to a certain limit at a particular instant of time and place due to heavy congestion of vehicles and increase of user base. The use of two frequency radar systems in high-end cars can be very helpful for seeing the path ahead during heavy fog. Introducing circular polarization in an antenna is an added advantage. LRR (long range radar) and SRR (short range radar) can help to avoid any possibility of collision of vehicles during low visibility conditions [9]. SRR and LRR operate on the principle of LIDAR (light detection and ranging) [10]. These are RF systems integrated in high-end cars. The real time status of restaurants, airports, cinemas, shopping complexes, traffic, train schedules, etc., can be known in the vehicle itself. Vehicles are equipped with remote monitoring of biomedical parameters of on board patients such as heart rate, blood pressure, EEG, ECG, etc. for remote monitoring [11,12].

Housing many RF systems inside the car cabin poses a problem of EMI (electromagnetic interference) and EMC (electromagnetic compatibility) [12]. The number of antennas used for different services in the car can be minimized using multi input multi output (MIMO) antennas [13]. Hence, isolation becomes an important parameter for RF systems [14]. The MIMO antenna has a good diversity of parameters, such as isolation, ECC (envelope correlation coefficient), DG (directive gain), MEG (mean effective gain) and CCL (coupling coefficient) or signal transfer rate. Multiband DRA is most suitable for vehicular services [15]. Reliable vehicular communication is a big challenge for autonomous cars.

14.2 DESIGN PROCESS OF APERTURE COUPLED DRA

DRA has low loss and high radiation efficiency. DRA can excite multiple modes simultaneously, and they are most stable. DRA material is available in different dielectric values, i.e., 10–100 permittivity. They can generate TE/TM/TEM modes. They are the best candidates for getting multi mode multiband features, which are requirements for vehicular wireless communication.

Step fix resonant frequency = 3.7 GHz, fix resonant mode (fundamental mode)

Impedance $50\,\Omega$ and substrate is given as substrate $FR4, \epsilon_r = 4.4$, height = 0.8 mm

λ_o Frequency space wavelength
λ_g Guided wavelength

Standard formulas used are

$$\lambda_o = \frac{C}{f_o} \tag{14.1}$$

$$\lambda_g = \frac{\lambda_o}{\sqrt{\epsilon_r \text{ effective}}}$$

Microstrip line: B and A are constants.

$$B = \frac{377\pi}{2\,z_o\sqrt{\varepsilon_r}} \quad A = \frac{20}{60}\sqrt{\frac{\varepsilon_r + 1}{2}} + \frac{\varepsilon_r - 1}{\varepsilon_r + 1}\left(0.23 + \frac{0.11}{\varepsilon_r}\right)$$

$$\frac{w}{d} = \left[\frac{8e^A}{e^{2A} - 2} \cdot \frac{2}{\pi}\left[B - 1\right] - \ln\left[2B - 1\right]\right] + \frac{\varepsilon_r - 1}{2\varepsilon_r}\left[\ln\left[B - 1\right] + 0.39 - \frac{0.61}{\varepsilon_r}\right]$$

$Z_o = 50\Omega$ input impedance of microstrip line (required for matching)

W is the width of microstrip line

d is the thickness of FR-4 (0.8 mm)

Width obtained for microstrip line

   * Length of microstrip line characteristic $= \lambda_o$

$$\text{Stub} = \frac{\lambda_g}{8}$$

$$L_s = \text{Slot} = 0.4\,\lambda_g$$

$$W_s = \text{Slot} = 0.2\,L_s\ (0.2 \times 0.4\,\lambda_g)$$

$\dfrac{W}{d} > 2$ (Aspect ratio mode dependent)

   Ground plane $= 4$ times of length

   (Infinite ground) $= 4$ times of width

   Finite ground plane $6h + L\ (Design\,DRA)$

$$6h + W\ (Design\,DRA)$$

14.3 COMPUTATION FOR 3.7 GHZ

Microstrip line:

$$B = \frac{377\pi}{2\,z_o\sqrt{\varepsilon_r}} = \frac{377 \times 3.14}{2 \times 50 \times \sqrt{4.4}} = 5.64$$

$$A = \frac{Z_o}{60}\sqrt{\frac{\varepsilon_r + 1}{2}} + \frac{\varepsilon_r - 1}{\varepsilon_r + 1}\left(0.23 + \frac{0.11}{\varepsilon_r}\right)$$

$$= \frac{50}{60}\sqrt{\frac{4.4 + 1}{2}} + \frac{4.4 - 1}{4.4 + 1}\left(0.23 + \frac{0.11}{4.4}\right) = 1.53$$

$$\frac{W}{d} = \left[\frac{8e^A}{e^{2A} - 2} \cdot \frac{2}{\pi}\,[B - 1] - \ln[2B - 1]\right] + \frac{\varepsilon_r - 1}{2\varepsilon_r}\left[\ln(B - 1) + 0.39 - \frac{0.61}{\varepsilon_r}\right]$$

$$= \left\{\frac{8e^{1.53}}{e^{2(1.53)} - 2} \cdot \frac{2}{\pi}\,[5.64 - 1] - \ln\left[2(5.64) - 1\right]\right.$$

$$\left. + \frac{4.4 - 1}{2(4.4)}\left[\ln[5.64 - 1] + 0.39 - \frac{0.61}{4.4}\right]\right\}$$

$$= (1.218)[4.64 - 2.33] + (0.386)[1.786]$$

$$\Rightarrow 3.5025$$

$$\frac{W}{d} = 3.5025, \ w = 3.5025 \times 0.8 \ \text{mm}$$

$$w = 2.8 \ \text{mm}$$

Stub:

$$f_r = 3.7 \, \text{GHz} \ , \lambda_o = 81 \ \text{mm} = \frac{C}{f_r}$$

$$\lambda_o = \frac{3 \times 10^8 \times 10^3}{3.7 \times 10^9} = 81 \ \text{mm}$$

$$\Rightarrow \lambda_g = \frac{\lambda_o}{\sqrt{\epsilon_r \ \text{effective}}} = \frac{81}{\sqrt{8.6}} = 27.62 \ \text{mm} \left[\varepsilon_{reffect} = \frac{4.4 + 12.8}{2} \right.$$

$$= 8.6$$

$$\Rightarrow \text{Length of stub} = \frac{\lambda_g}{8} = \frac{27.62}{8} = 3.4525 \ \text{mm}$$

Slot:

$$L_s = 0.4 \ \lambda_g = 0.4 \times 27.62$$

$$= 11.048 \ \text{mm}$$

$$W_s = 0.2 l_s = 0.2 \times 11.048 = 2.2096 \ \text{mm}$$

DRA dimensions: (TE_{111} fundamental mode)
Formula used

$$f_r = \frac{c}{2\pi \sqrt{\epsilon_r}}, \sqrt{\left(\frac{m\pi}{a}\right)^2 + \left(\frac{n\pi}{b}\right)^2 + \left(\frac{p\pi}{2d}\right)^2}$$

$$a = 20 \ \text{mm}$$

$$b = 20 \ \text{mm}$$

$$d = 14 \ \text{mm}$$

$$m = n = p = 1 \qquad \qquad \frac{a}{b} = \textbf{0.5 to 2.5}$$

$$\textbf{fix} \, d = 14 \ \text{mm} \qquad \qquad \frac{b}{d} = \textbf{0.5 to 2.5}$$

14.4   DRA DESIGN FORMULATIONS

$\epsilon_r k_0^2 = k_x^2 + k_y^2 + k_z^2$; characteristic equation.

$$k_x = m\pi/a$$

$$k_y = n\pi/b$$

$$k_z = p\pi/d$$

where a, b, d are dimensions and m, n, p are the indices.

$$k = 2\pi / \lambda = \omega\sqrt[3]{\mu\epsilon} = \omega/c;$$

$k_z\tan(k_z d/2) = \sqrt{(\epsilon r - 1)k_0^2 - k_z^2}$; **transcendental equation.**
$\int E^2\, dV = \int H^2\, dV$; **electrical energy and magnetic energy in DRA.**
Boundary conditions, PEC

$$n \times E = 0; \; n.H = 0;$$

The tangential component of electric field and normal component of magnetic field is equal to **"zero"** at the interface of DRA, $z = 0, d$.
   Boundary conditions, PMC

$$n \times H = 0; \; n.E = 0;$$

$$E_x = \frac{1}{j\omega\epsilon\left(1+\dfrac{\gamma^2}{k^2}\right)}\left[\frac{\partial H_z}{\partial y} - \frac{1}{j\omega\mu}\frac{\partial^2 E_z}{\partial z\partial x}\right] \qquad E_y = \frac{1}{j\omega\epsilon\left(1+\dfrac{\gamma^2}{k^2}\right)}\left[-\frac{1}{j\omega\mu}\frac{\partial^2 E_z}{\partial z\partial y} - \frac{\partial H_z}{\partial x}\right]$$

$$H_x = \frac{-1}{j\omega\mu\left(1+\dfrac{\gamma^2}{k^2}\right)}\left[\frac{\partial E_z}{\partial y} - \frac{1}{j\omega\epsilon}\frac{\partial^2 H_z}{\partial z\partial x}\right]$$

$$H_y = \frac{-1}{j\omega\mu\left(1+\dfrac{\gamma^2}{k^2}\right)}\left[\frac{1}{j\omega\epsilon}\frac{\partial^2 H_z}{\partial z\partial y} - \frac{\partial E_z}{\partial x}\right]$$

TE mode $\left(E_z = 0 \text{ and } H_z \neq 0\right)$

$$H_z = B\sin\left(\frac{m\pi}{a}x\right)\sin\left(\frac{n\pi}{b}y\right)\sin(k_z z)$$

Similarly, E_z fields can be evaluated for TM mode.

14.5 DESIGN DEVELOPMENT AND EVALUATION OF DRA

DRA has been tested for antenna parameters at multiple bands. A 40 GHz VNA is used to measure S_{11}, as shown in Figure 14.1. The HFSS model is also depicted in Figures 14.2 and 14.3. An equivalent circuit model is shown in Figure 14.4 with Zin solution. Figure 14.5 shows a sketch of aperture coupled DRA. Design dimensions of DRA are placed in Table 14.1. E-fields are shown in Figures 14.6 and 14.7. H-fields are shown in Figures 14.8 and 14.9.

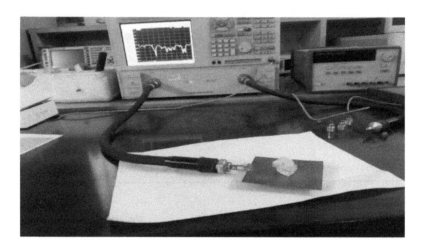

FIGURE 14.1 Hardware of aperture coupled DRA at 3.6 GHz, measurements by VNA.

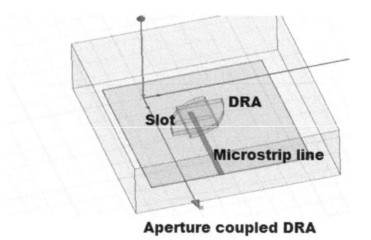

Aperture coupled DRA

FIGURE 14.2 HFSS model of aperture coupled DRA at 3.6 GHz.

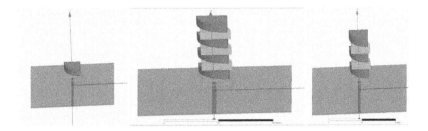

FIGURE 14.3 HFSS model of aperture coupled DRAs for vehicle.

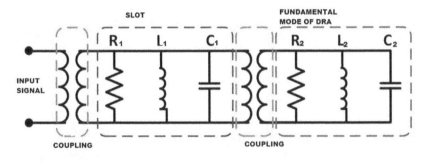

FIGURE 14.4 DRA equivalent circuit.

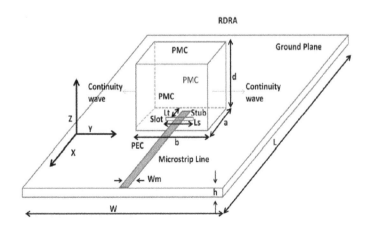

FIGURE 14.5 DRA as boundaries value problem.

TABLE 14.1
Vehicle Services and Spectral Frequencies

Service	Typical Frequency
AM Radio	Approximately 1 MHz
FM Radio	88–108 MHz
In-vehicle VHF-TV	50–400 MHz
Digital Audio Broadcasting (DAB)	100–400 MHz
Remote Keyless Entry (RKE)	315/413/434 MHz
Tire Pressure Monitoring System (TPMS)	315/413/434 MHz
Cellular Phone, GSM	850, 900, 1800, 1900 MHz, (2100 MHz)
IMT (International Mobile Telephony)	2300–2400 MHz
	2700–2900 MHz
	3400–4200 MHz
	4400–4900 MHz
Satellite Navigation (GPS)	1.575 GHz
WLAN	2400–2485 MHz
	5150–5350 MHz
	5725–5825 MHz
Satellite Digital Audio Radio Service (SDARS)	2.3 GHz
Wi-Fi	2.4/5.8 GHz
Bluetooth	2.4 GHz
Wi-Max	2.3/2.5/3.5 GHz
	2500–2690 MHz
	3400–3690 MHz
	5250–5850 MHz
Electronic Toll Collection (ETC)	5.8 GHz (or 900 MHz)
V2V (Vehicle to Vehicle)	5.9 GHz
Collision Avoidance Radar (LRR & SRR)	24 and 77 GHz
5G Communication	24.5–27.5 GHz and 1–6 GHz

Let R_1, L_1, C_1, $R2$, $L2$, and $C2$ are taken as R_s, L_s, C_s, R_D, L_D, C_D respectively.

$$Y_D = \frac{1}{R_D} + \frac{1}{s\,L_D} + s\,C_D$$

$$Y_s = \frac{1}{R_s} + \frac{1}{sL_s} + s\,C_S$$

$$Z_{\text{in}\,D} = \frac{1}{n^2}\,Z_D = \frac{1}{n^2}\frac{1}{Y_D} = \frac{1}{n^2\left(\dfrac{1}{R_s} + \dfrac{1}{s\,L_s} + s\,C_S\right)}$$

$$Z_{\text{in}D} = \frac{s\,L_s\,R_s}{n^2\left(R_s + s\,L_s + s^2 L_s C_s R_s\right)}$$

$$Z_s = \frac{1}{Y_s} + Z_{inD}$$

$$Z_s = \frac{1}{\dfrac{1}{R_s} + \dfrac{1}{sL_s} + sC_s} + Z_{in}$$

$$Z_s = \frac{R_s\, sL_s}{s^2 L_s\, C_s\, R_s + R_s + sL_s} + \frac{sL_D R_D}{n^2\left(R_D + sL_D + s^2\, L_D C_D R_D\right)}$$

$$Z_{in} = \frac{1}{n^2} Z_{inD}$$

$$Z_{in} = \frac{1}{n^4}\Big[s^3\left(n^2 L_D R_D C_D R_s L_s + L_D R_D L_s R_s C_s\right) + s^2\left(n^2 L_D L_s R_s + L_D R_D R_s\right)$$

$$+ s\left(L_D R_D R_s + R_s R_D L_s\right)\Big]\Big/$$

$$\Big[\left(s^4 L_D\ R_D\ C_D L_s R_s\ C_s\right) + s^3\left(L_D R_D C_D L_s + L_D L_s R_s\ C_s\right)$$

$$+ s^2\left(L_D R_D C_D R_D + L_s R_s\ C_s R_s\right) + S\left(L_D R_s + L_s R_D\right) + R_s R_D\Big]$$

$$\epsilon_r\, ko^2 = kx^2 + ky^2 + kz^2;$$

where wave vector $(k) = \omega\sqrt{\mu\epsilon}$
Where wave vector in x, y and z directions are $kx = \frac{m\pi}{a}$; $ky = \frac{n\pi}{b}$; $kz = \frac{p\pi}{d}$; $k_z \tan\left(k_z d / 2\right) = \sqrt{\left(\epsilon_r - 1\right)k_0^2 - k_z^2}$; from transcendental equation.
Ground plane dimension requirements are (Lg × Wg):

$$Lg = L + 6\,h;$$

where h is the height of substrate, $Wg = W + 6h + 6h$;
$\lambda = \frac{c}{f}$; operating wavelength of antenna;
Permitivity $r_{eff} = \left(\frac{\lambda_o}{\lambda_g}\right)^2$; λ_o, λ_g are free space and guide wavelengths.
Resonant frequency of DRA is calculated by the formulation given below:

$$\epsilon_r = \frac{\epsilon}{\epsilon_o};\ fr_{mnp} = \frac{c}{2\pi\sqrt{\epsilon r}}\sqrt{\left(\frac{m\pi}{a}\right)^2 + \left(\frac{n\pi}{b}\right)^2 + \left(\frac{p\pi}{2d}\right)^2}$$

Slot length, slot width and stub dimension of DRA formulations are given as:

$$L_S = \frac{0.4\lambda_o}{\sqrt{\varepsilon_e}};\ W_s = 0.2L_s;\ L_{stub} = \frac{\lambda_g}{4}$$

$\frac{\lambda}{2\pi} < r < \frac{2D^2}{\lambda}$; near field relation

$k \gg 1$; far field relation as $r > \frac{2D^2}{\lambda}$;

where λ_o is the wavelength and the effective permittivity is defined as:

$$\varepsilon_{\text{effective}} = \frac{\varepsilon_r + \varepsilon_s}{2}$$

where ε_r and ε_s are the dielectric constant of the rectangular dielectric resonator and substrate, respectively.

$$\lambda_g = \frac{\lambda_0}{\sqrt{\varepsilon_{\text{eff}}}}$$

where λ_0, λ_g are free space and guided wavelengths.

Boundary conditions generally depend upon PEC (electric walls) and PMC (magnetic walls). For PEC walls, $n \times E = 0$ and $n.H = 0$. For PMC walls, $n \times H = 0$ and $n.E = 0$.

DRA operates at 3.5 GHz for Wi-Max, 5.1 GHz for WLAN and 8.5 GHz for X band radars.

DRA dimensions are given in Table 14.1 and the results obtained by simulation and experimentation on S-parameters, E- and H-fields and radiation patterns are given in Figures 14.6–14.13 (Table 14.2).

14.6   DRA EVALUATION

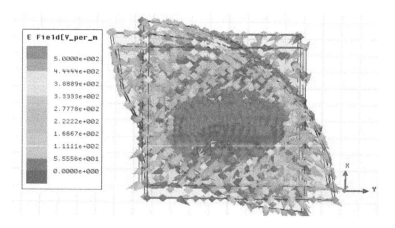

FIGURE 14.6   E-field vectors inside DRA.

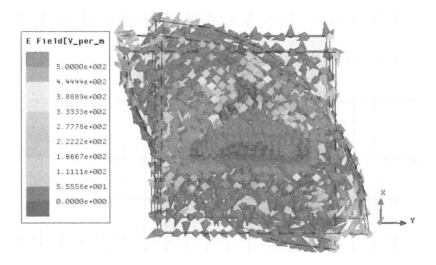

FIGURE 14.7 E-field vectors inside DRA.

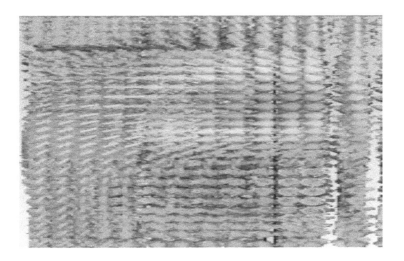

FIGURE 14.8 H-field vector inside DRA.

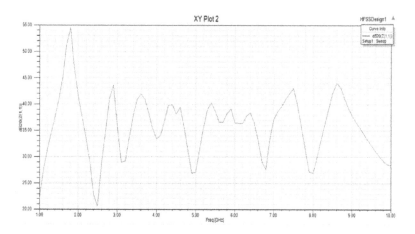

FIGURE 14.9 Z_{11} of DRA.

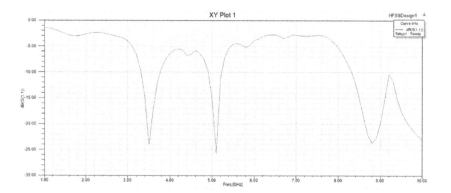

FIGURE 14.10 S_{11} of DRA below −10 dB.

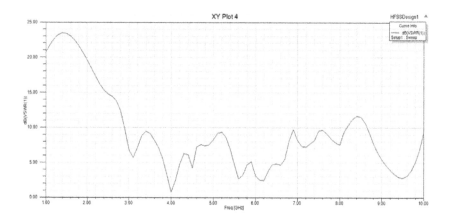

FIGURE 14.11   VSWR of DRA below 2 dB.

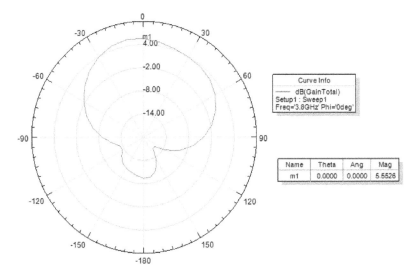

FIGURE 14.12 Radiation pattern of DRA more than 4 dB at broad side.

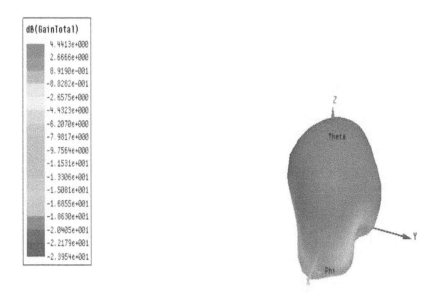

FIGURE 14.13 3D radiation pattern of DRA.

TABLE 14.2
DRA Design Dimensions

Nomenclature	mm (dimensions)
Microstrip line length	65
Microstrip line width	32
Ground plane	110×100
Height of substrate	1.6
Slot length	15
Slot width	3
Stub length	25
Stub width	15
DRA (each slab)	$20 \times 20 \times 7$

14.7 CONCLUSION

DRA antennas have been designed and developed for vehicular applications for Wi-Fi applications at 3.6 GHz. Mathematical background has been developed to evaluate for design radiated fields and frequency of DRA. Multiband DRA is most suitable for vehicular applications. A maximum number of services can be obtained from one single DRA if it can resonate at multiple frequencies. MIMO antenna also meet most of the requirements.

REFERENCES

1. S. Sun, T. S. Rappaport, M. Shafi, P. Tang, J. Zhang, and P. J. Smith, "Propagation Models and Performance Evaluation for 5G Millimeter-Wave Bands," *IEEE Trans. Veh. Technol.*, vol. 67, no. 9, pp. 8422–8439.
2. J. Medbo, P. Kyosti, K. Kusume, L. Raschkowski, K. Haneda, T. Jamsa, V. Nurmela, A. Roivainen, and J. Meinila, "Radio Propagation Modeling for 5G Mobile and Wireless Communications," *IEEE Commun. Mag.*, vol. 54, no. 6, pp. 144–151, 2016.
3. S. Hur, T. Kim, D. J. Love, J. V. Krogmeier, T. A. Thomas, and A. Ghosh, "Millimeter Wave Beamforming for Wireless Backhaul and Access in Small Cell Networks," *IEEE Trans. Commun.*, vol. 61, no. 10, pp. 4391–4403, 2013.
4. C. E. Zebiri, M. Lashab, D. Sayad, I. T. E. Elfergani, K. H. Sayidmarie, F. Benabdelaziz, R. A. Abd-Alhameed, J. Rodriguez, and J. M. Noras, "Offset Aperture-Coupled Double-Cylinder Dielectric Resonator Antenna with Extended Wideband," *IEEE Trans. Antenn. Propag.*, vol. 65, no. 10, pp. 5617–5622, 2017.
5. R. S. Yaduvanshi and H. Parthasarathy, *Rectangular Dielectric Resonator Antennas Theory and Design*. New Delhi/Heidelberg/New York/Dordrecht/London: Springer, 2016.
6. G. Varshney, V. S. Pandey, R. S. Yaduvanshi, and L. Kumar, "Wide Band Circularly Polarized Dielectric Resonator Antenna with Stair-Shaped Slot Excitation," *IEEE Trans. Antenn. Propag.*, vol. 65, no. 3, pp. 1380–1383, 2016.
7. M. Singh, R. S. Yaduvanshi, A. Vaish, "Design for Enhancing Gain in Multimodal Cylindrical Dielectric Resonator Antenna," *IEEE India Conference*, December 2015.

8. S. Gotra, G. Varshney, R. S. Yaduvanshi, and V. S. Pandey, "Dual-Band Circular Polarisation Generation Technique with the Miniaturization of a Rectangular Dielectric Resonator Antenna," *IET Microw. Antenn. Propag.*, vol. 13, no. 10, pp. 1742–1748, 2019.

9. M. Singh, A. K. Gautam, R. S. Yaduvanshi, and A. Vaish, "An Investigation of Resonant Modes in Rectangular Dielectric Resonator Antenna Using Transcendental Equation," *Wireless Pers. Commun.*, vol. 95, no. 3, pp. 2549–2559, 2017.

10. G. Bakshi, A. Vaish, and R. S. Yaduvanshi, "Two-Layer Sapphire Rectangular Dielectric Resonator Antenna for Rugged Communications," *Progr. Electromag. Res. Lett.*, vol. 85, pp. 73–80, 2019.

11. G. Kumar, M. Singh, S. Ahlawat, and R. S. Yaduvanshi, "Design of Stacked Rectangular Dielectric Resonator Antenna for Wideband Applications," *Wireless Pers. Commun.*, vol. 109, no. 3, pp. 1661–1672, 2019.

12. R. Khan, M. H. Jamaluddin, J. U. R. Kazim, J. Nasir, and O. Owais, "Multiband-Dielectric Resonator Antenna for LTE Application," *IET Microw. Antenn. Propag.*, vol. 10, no. 6, pp. 595–598, 2016.

Annexure A1: Case Studies and Prototype DRAs

GAIN, BANDWIDTH, POLARIZATION, RADIATION PATTERN AND BEAM WIDTH CONTROL IN DRA

Most of the hardware discussed in this annexure has been developed in the lab. Simulated and prototype models have been depicted. Their dimensions along with results are presented as case studies. Dielectric resonator antenna (DRA) has large advantages as compared to bandwidth due to the fact that antenna parameters can be easily controlled. A large number of resonant modes can be easily excited due to flexibility available to antennas designer due to permittivity and aspect ratio options.

Case 1 Four Port MIMO DRA (Figure A1.1)

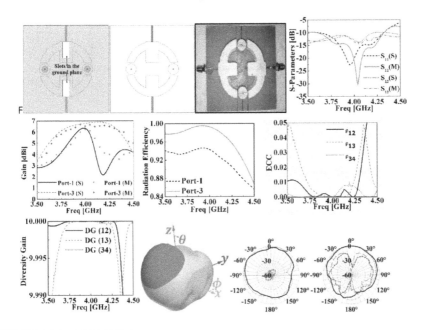

FIGURE A1.1   MIMO DRA.

Case 2 Multiple Modes Excited into DRA (Figures A1.2 and A1.3)

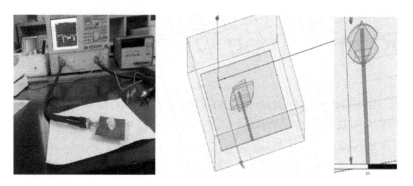

FIGURE A1.2   (a) Hardware model and (b) HFSS model.

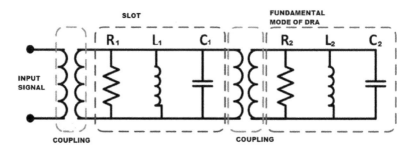

FIGURE A1.3   Equivalent circuit of aperture coupled DRA.

The quality factor of DRA can be obtained by going to HFSS modeler eigen mode, keeping it as isolated DRA with double of height and removing input excitation. This way quality factor is obtained, which can be compared with *R, L, C* circuit quality factor, and the possible values of equivalent circuit can thus be obtained. The eigen values in HFSS are without excitation. Equivalent circuits of DRA can be drawn and verified with the MATLAB program.

MATLAB Program: DRA MATLAB program

```
close all;
clear all;
w=10000:1:10000000;
N1=1/0.92;
N2=1/1.2;
Rs=955;
Rd1=290;
Rd2=207;
Ls=0.419e-9;
Ld1=1.38e-9;
Ld2=3.193e-9;
```

```
Cs=0.66e-12;
Cd1=2.78e-12;
Cd2=1.8e-12;
Lc=1.86e-9;
Cc=0.86e-12;

% N1=input('enter the value of N1=');
% N2=input('enter the value of N2=');
% Rs=input('enter the value of Rs=');
% Ls=input('enter the value of Ls=');
% Cs=input('enter the value of Cs=');
% Rd1=input('enter the value of Rd1=');
% Ld1=input('enter the value of Ld1=');
% Cd1=input('enter the value of Cd1=');
% Rd2=input('enter the value of Rd2=');
% Ld2=input('enter the value of Ld2=');
% Cd2=input('enter the value of Cd2=');
% Lc=input('enter the value of Lc=');
% Cc=input('enter the value of Cc=');
x1=w*Cs;
x2=1./(w*Ls);
x3=Rs;
x4=w*Cd1;
x5=1./(w*Ld1);
x6=Rd1;
x7=w*Cd2;
x8=1./(w*Ld2);
x9=Rd2;
x10=x6./(((x6)^2)*(x4-x5).^2);
x11=x9./(((x9)^2)*(x7-x8).^2);
p=x10+x11;
x12=w*Lc;
x13=1./(w*Cc);
x14=x12-x13;
x15=(x6)^2*(x4-x5)./(1+(x4-x5).^2);
x16=(x9)^2*(x7-x8)./(1+(x7-x8).^2);
q=x14-x15-x16;

% y=sqrt(N1^2*((1/x3)+p*N2^2./(p.^2+q.^2))).^2+(N1^2*(x1-x2)-
(q*N2^2./(p.^2+q.^2)).^2);
x17=(N1^2*((1/x3)+p*N2^2./(p.^2+q.^2))).^2;
x18=(N1^2*((x1-x2)-N2^2*q./(p.^2+q.^2)).^2);
y=sqrt(x17.^2+x18.^2);
 z=1./y;
 s= (x12+x4./((1/x6^2)+(x4-x5).^2)+x7./((1/x9^2)+(x7-x8).^2));
 x19=N1^2*x1+(N1^2*N2^2./(p.^2+q.^2)).*s;
 x20=N1^2*(1/x3+p*N2^2./(p.^2+q.^2));
 g=x19./x20;
 x21=sqrt((p-sqrt(Ls/Cs)).^2+(q+Rs./(2*w*sqrt(Ls*Cs))).^2);
 x22=sqrt((p+sqrt(Ls/Cs)).^2+(q-Rs./(2*w*sqrt(Ls*Cs))).^2);
 h=x21./x22;
```

```
subplot(2,2,1)
 plot(w,z, 'r')
 title("INPUT IMPEDANCE");
xlabel("W");
ylabel("IMPEDANCE");

%    subplot(2,2,2)
%     plot(w,y, 'b')
%      title("INPUT ADMITANCE");
%    xlabel("W");
%    ylabel("ADMITANCE");
 subplot(2,2,3)
  plot(w,h)
  title(" REFLECTION COFFICIENT");
 xlabel("W");
 ylabel("S11");
%  plot(w,x17,'b')
 subplot(2,2,4)
  plot(w,g)
  title("QUALITY FACTOR");
 xlabel("W");
 ylabel("IMPEDANCE");
%  plot(w,x18,'r')
```

(Figures A1.4–A1.13)

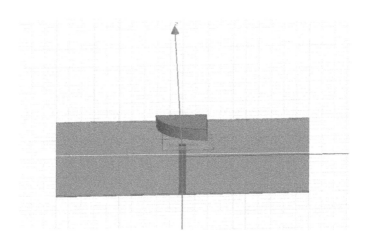

FIGURE A1.4   DRA two layer.

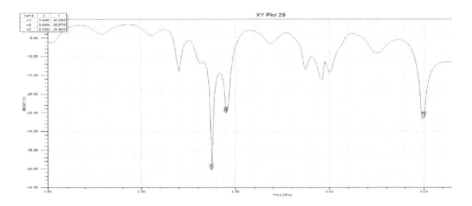

FIGURE A1.5   DRA two layer reflection coefficient.

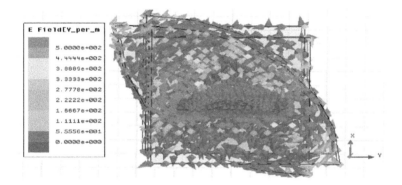

FIGURE A1.6   E-field vector diagram.

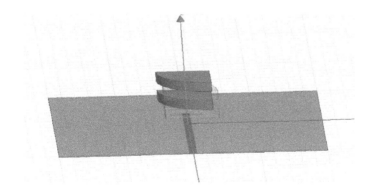

FIGURE A1.7   DRA four layers.

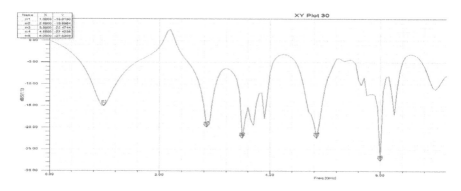

FIGURE A1.8 DRA four layers reflection coefficient.

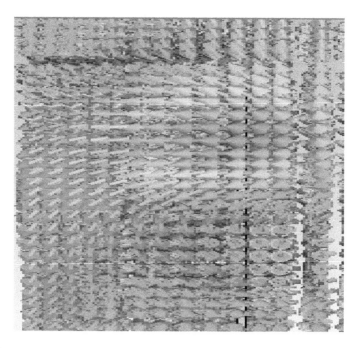

FIGURE A1.9 Higher order mode.

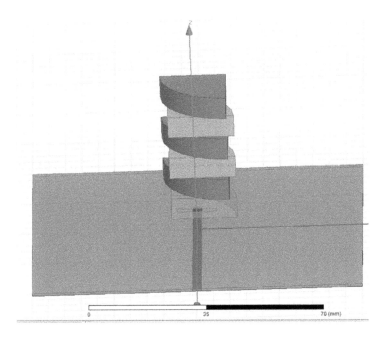

FIGURE A1.10   DRA eight layers.

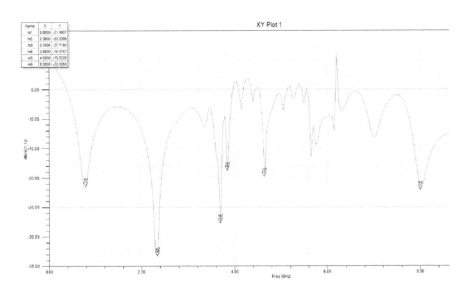

FIGURE A1.11   DRA eight layers reflection coefficient.

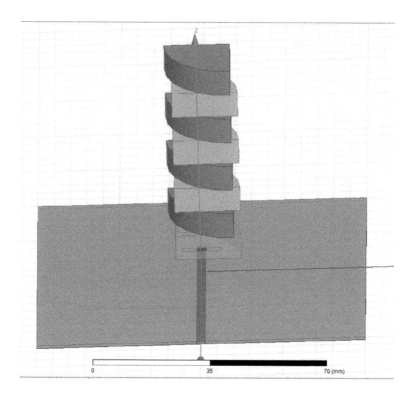

FIGURE A1.12   DRA ten layers.

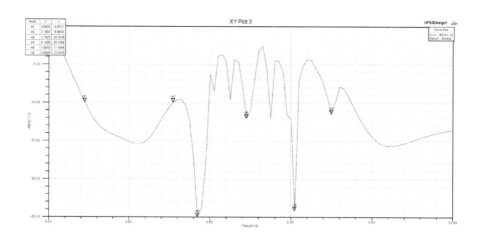

FIGURE A1.13   DRA ten layers reflection coefficient.

Case 3 Patterned DRA (Figures A1.14–A1.23)

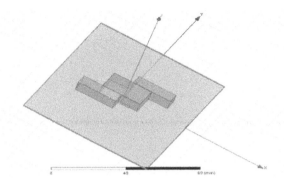

FIGURE A1.14   DRA HFSS model.

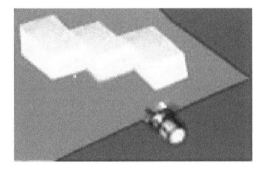

FIGURE A1.15   Hardware model.

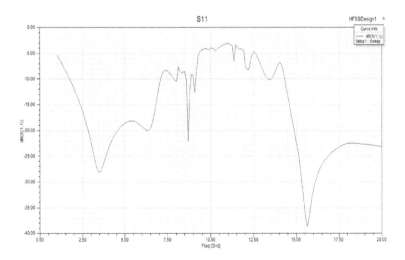

FIGURE A1.16 S_{11} plot of pattern DRA.

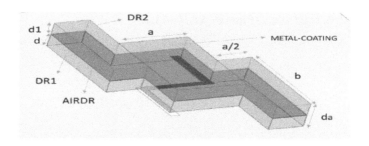

FIGURE A1.17 HFSS model.

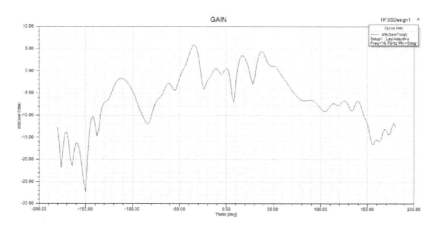

FIGURE A1.18 Gain plot of pattern DRA.

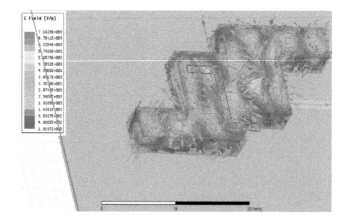

FIGURE A1.19 E-fields.

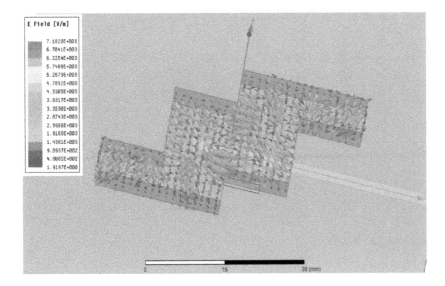

FIGURE A1.20   E-fields.

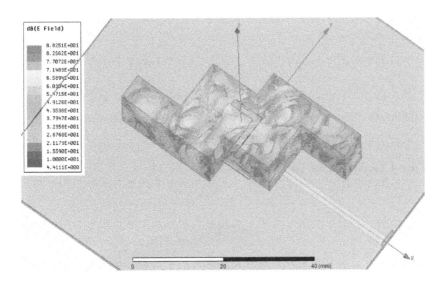

FIGURE A1.21   E-fields.

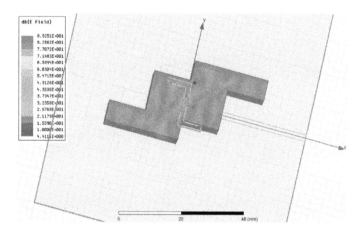

FIGURE A1.22   E-fields.

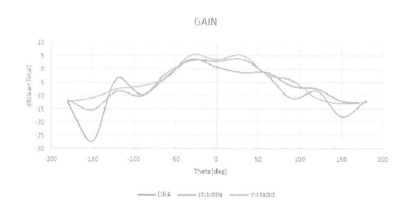

FIGURE A1.23   Comparative gain of DRA, stack DRAs and multiple stack DRAs.

Case 4 Trishul Shape DRA (Figures A1.24–A1.27)

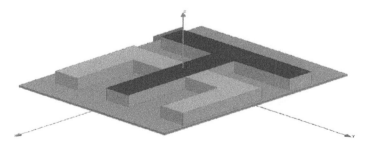

FIGURE A1.24   HFSS model.

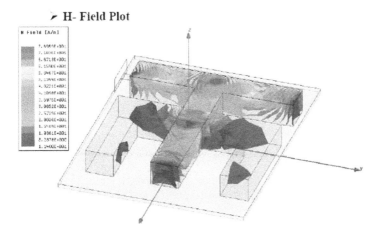

FIGURE A1.25   H-fields.

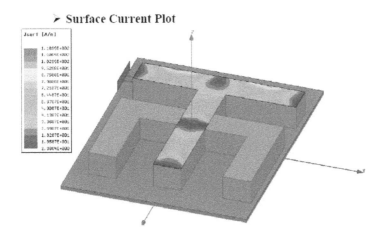

FIGURE A1.26   E-field.

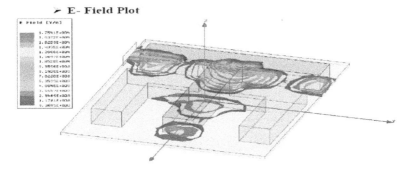

FIGURE A1.27   E-fields.

Case 5 RF Liquid Sensor: Milk, Water, Oil and Other Liquids (Figures A1.28–A1.30)

FIGURE A1.28   Hardware model.

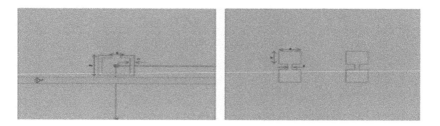

FIGURE A1.29   HFSS model.

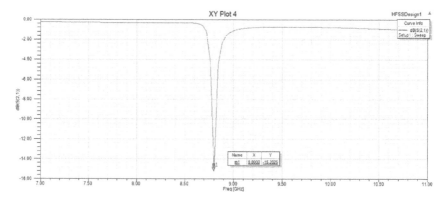

FIGURE A1.30   Reflection coefficient.

Case 6 Circular Polarization in DRA (Figures A1.31–A1.33)

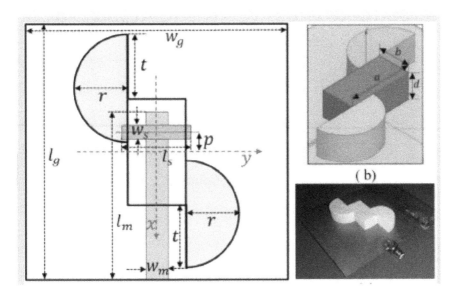

FIGURE A1.31   Hardware and HFSS models.

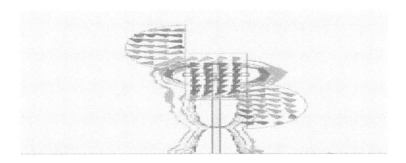

FIGURE A1.32   E-fields.

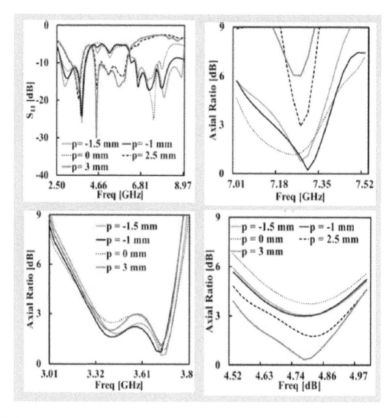

FIGURE A1.33   Reflection coefficient (S_{11}) and Axial Ratio plots.

Case 7 Circular Polarization in DRA (Figures A1.34 and A1.35)

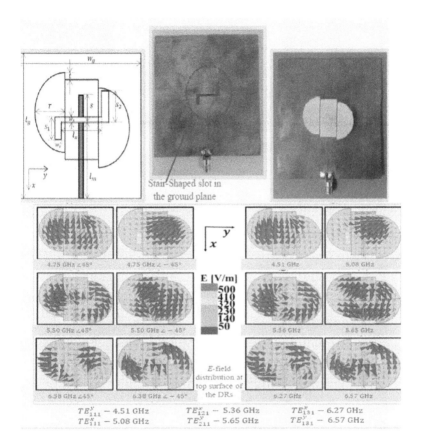

$$TE_{111}^y - 4.51 \text{ GHz} \qquad TE_{121}^x - 5.36 \text{ GHz} \qquad TE_{131}^y - 6.27 \text{ GHz}$$
$$TE_{111}^x - 5.08 \text{ GHz} \qquad TE_{211}^y - 5.65 \text{ GHz} \qquad TE_{131}^y - 6.57 \text{ GHz}$$

FIGURE A1.34 Circular polarized DRA.

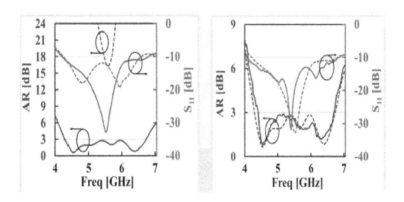

FIGURE A1.35 Reflection coefficient (S_{11}) and Axial Ratio plots.

Case 8 Vehicular DRA with Multiband Circular Polarization (Figure A1.36)

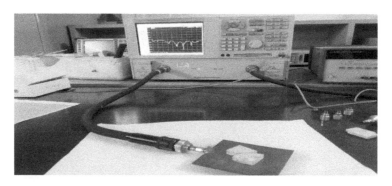

FIGURE A1.36   VNA measurement set up for s-parameters.

Case 9 RDRA (Figure A1.37)

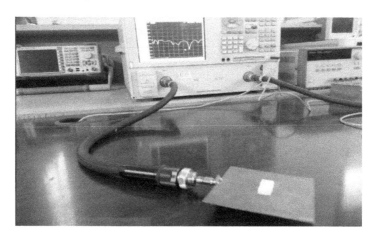

FIGURE A1.37   VNA measurement set up for s-parameters.

Case 10 Rectangular DRA with Slits (Figure A1.38)

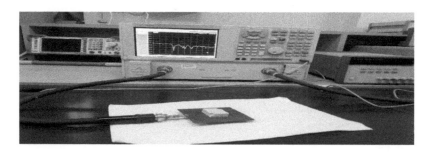

FIGURE A1.38   VNA measurement set up for s-parameters.

Case 11 Stacked DRA (Figure A1.39)

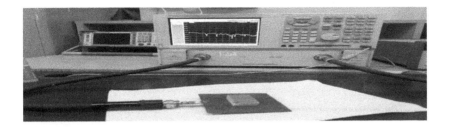

FIGURE A1.39   VNA measurement set up for s-parameters.

Case 12 Multiband DRA (Figure A1.40)

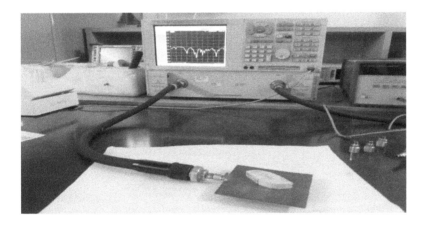

FIGURE A1.40   VNA measurement set up for s-parameters.

Case 13 Multiband DRA (Figure A1.41)

FIGURE A1.41   VNA measurement set up for s-parameters.

Case 14 Sapphire DRA (Figure A1.42)

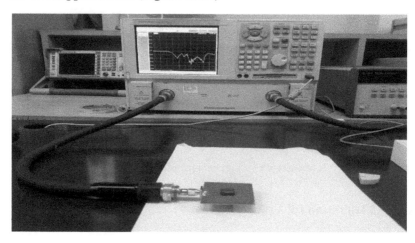

FIGURE A1.42   VNA measurement set up for s-parameters.

Case 15 Rectangular Stacked DRA (Figure A1.43)

FIGURE A1.43   VNA measurement set up for s-parameters.

Case 16 Rectangular Horn DRA (Figure A1.44)

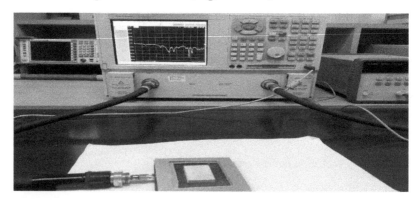

FIGURE A1.44   VNA measurement set up for s-parameters.

Case 17 DRA in Stacking for Increasing Bandwidth (Figure A1.45)

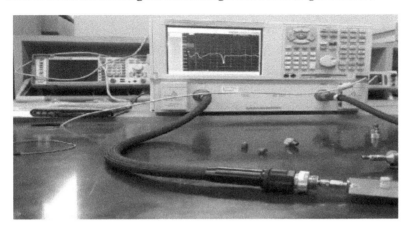

FIGURE A1.45   VNA measurement set up for s-parameters.

Annexure A2: Generation of Resonant Modes and Control

Dielectric wave guide modeling (DWM) has been used to predict the resonant frequencies f_{pmn} of the modes $TE^x_{\delta mn}$ generated in the dielectric rectangular resonator. Direction of propagation has been assumed in the x-axis. Merging or shifting, separation, generation of even and odd and hybrid modes are today's DRA science to get wide band, multibands, high gain, narrow beam, etc., in single DRA. Hence, manipulation of modes can provide different applications in a single DRA.

$$k_x \sec\left(k_x \frac{d}{2}\right) = \sqrt{(\varepsilon_r - 1)}\, k_{mn};$$

where:

$$k_{mn} = \frac{2\pi f_{mn}}{c}, k_y = \frac{m\pi}{b}, k_z = \frac{n\pi}{2h}.$$

$$k_x^2 + k_y^2 + k_z^2 = \varepsilon_r k_{mn}^2; \text{ and } c = 3\times10^8 \text{ m/s}$$

Thus, e. m. wave fields H_z or E_z can be expressed as propagating fields as given below:

$$E_z\left(x,y,z,t\right) = \sum_{m,n} u_{m,n}\left(x,y\right)\mathrm{Re}\left(C_{m,n}\ e^{j(\omega t - \gamma_{m,n}z)}\right)$$

where $u_{m,n}\left(x,y\right) = \dfrac{2}{\sqrt{ab}}\sin\left(\dfrac{m\pi x}{a}\right)\sin\left(\dfrac{n\pi y}{b}\right)$; when $k \geq h_{m,n}$ propagating modes.

In this equation, $C_{m,n}$ are the coefficients, and the wave is propagating in z direction.

$$\therefore\ \gamma_{m,n}\left(\text{Propagation constant}\right) = \sqrt{h_{m,n}^2 - k^2} = \sqrt{h_{m,n}^2 - \omega^2 \mu\epsilon}$$

where $h_{m,n} = k_c = n\pi/b$ are possible eigen values.

The computation of field H_z or E_z shall mostly depend upon DRA boundary conditions. If all six walls of DRA transparent i.e., magnetic walls (PMC walls) fields will be different from DRA with top and bottom as PEC (electrical walls) and rest four are PMC (magnetic walls). We are well versed that $H_z = 0$ at magnetic walls and $E_z = 0$ at electric walls. Here, $n \times E = 0$ and $n. H = 0$ for the PEC case and $n \times H = 0$ and $n. E = 0$ for the PMC case.

Hence, E_z, is electric field component:

$$E_z\left(x,y,z,t\right) = \sum_{m,n} u_{m,n}\left(x,y\right)\mathrm{Re}\left(C_{m,n}\ e^{j(\omega t - \gamma_{m,n}z)}\right)$$

where $u_{m,n}(x,y) = \frac{2}{\sqrt{ab}} \sin\left(\frac{m\pi x}{a}\right) \sin\left(\frac{n\pi y}{b}\right)$; when $k \geq h_{m,n}$ propagating modes.

$C_{m,n}$ Fourier coefficients of modes:

Amplitude coefficient:

$$C_{m,n} = \frac{j\omega\mu I\,dl}{4\pi} \int_o^a \int_o^b \frac{u_{m\,n(x,y)}}{\sqrt{\left(x-\frac{a}{2}\right)^2 + \left(y-\frac{b}{2}\right)^2}}\, e^{-jk\sqrt{\left(x-\frac{a}{2}\right)^2 + \left(x-\frac{b}{2}\right)^2}}\, dx\,dy;$$

Resonant frequency: $\pi^2 \left[\left(\frac{m}{a}\right)^2 + \left(\frac{n}{b}\right)^2\right] \leq \omega^2$

$$\gamma(mn) = \frac{j\pi p}{d}$$

$$k^2 + \gamma_{mn}^2 = h_{mn}^2$$

hence, $k^2 = h_{mn}^2 + \frac{\pi^2 p^2}{d^2}$

where:

u_{mn} depends on input excitation,

h_{mn} resonant mode (cut-off frequency), and

k propagation constant generation of modes or characteristics frequencies $\omega(mnp)$ of $e \cdot m$. Field oscillations inside the cavity resonator have been discussed. The basic Maxwell's theory can be applied with boundary conditions to express resonator fields as superposition of these characteristic frequencies.

$$E_z(x,y,z,t) = \sum_{m,n,p} \text{Re}\int C_{m,n,p}\, e^{j\omega(mnp)t}\, u_{mnp}(x,y,z)$$

Or

$$\sum_{m,n,p} |C_{mnp}| u_{mnp}(x,y,z) \cos\left(\omega(mnp) + \phi(mnp)\right);$$

where $u_{m,n}(x,y) = \frac{2}{\sqrt{ab}}\sin\left(\frac{m\pi x}{a}\right)\sin\left(\frac{n\pi y}{b}\right)$; $\omega(mnp)$ is the characteristic frequency and $\phi(mnp)$ is the phase of current applied. The rectangular cavity resonator is excited at the center with an antenna probe carrying current $I(t)$ of some known frequency $\omega(mnp)$. This generates the field E_z inside the cavity of the form given as follows:

$$E_z(x,y,\delta,t)$$

$$= \int G(x,y) \frac{j\omega\mu I\,dl\left(x^2+y^2\right)}{4\pi\left(x^2+y^2+\delta^2\right)^{3/2}}\, e^{\left(j\omega t - \frac{\omega}{c}\sqrt{x^2+y^2+\delta^2}\right)} \cdot I(\omega)e^{jkt}\, d\omega$$

where $G(x, y)$ are constant terms associated with current.

Equating resonator current fields with the antenna current fields at $z = \delta$ plane is expressed as:

$$\sum_p |Cmnp| \sqrt{\tfrac{2}{d}} \sin\left(\tfrac{p\pi\delta}{d}\right) \cos(\omega(mn\,p)t + \phi\,(mn\,p)) \text{ will be equal } \int G(x, y)$$

$$\frac{j\omega\,\mu I\,dl(x^2+y^2)}{4\pi\,(x^2+y^2+\delta^2)^{3/2}}\, I(\omega)e^{jkt}\, d\omega\,(e^{\left(j\omega t - \frac{\omega}{c}\sqrt{x^2+y^2+\delta^2}\,+\psi_{mn\,p}\right)}\, u_{m,n}\,(x, y)\mathrm{d}x\,\mathrm{d}y;\text{ probe input case.}$$

It is clear that for these two expressions to be equal, the probe current can be defined as follows:

$$I(\omega) = \frac{1}{2} \sum_{m,n\rho} |I(mn\,p)| \left[\delta(\omega - \omega(mnp))e^{j\varnothing(mnp)} + e^{j\varnothing(mn\,p)}\,\delta(\omega + \omega\,(mn\,p))\right]$$

The antenna probe current must contain only the resonator characteristics frequencies $\omega(mn\,p)$, then

$$\sum_p |Cmnp| \sqrt{\frac{2}{d}} \sin\left(\frac{p\pi\delta}{d}\right) \cos(\omega(mn\,p)t + \phi(mn\,p)$$

$$= \int G(x, y) \frac{j\omega\,\mu I\,dl(x^2+y^2)}{4\pi(x^2+y^2+\delta^2)^{3/2}}\, .I(\omega)e^{jkt}\, d\omega$$

$$e^{\left(j\omega t - \frac{\omega}{c}\sqrt{x^2+y^2+\delta^2}\,+\psi_{mn\,p}\right)}\, u_{m,n}(x, y)\quad \mathrm{d}x\,\mathrm{d}y;$$

Antenna input current = DRA radiated output current

Hence, we can conclude that modes in the cavity resonator shall mostly depend on the type of excitation, point of excitation, size or dimensions of cavity and coupling coefficient of environment. Figures A2.1 through A2.4 are formats of E-fields established due to input excitation of DRA. These are called resonant modes due to E-field vectors.

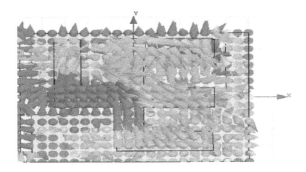

FIGURE A2.1   DRA E-fields.

FIGURE A2.2   DRA E-fields.

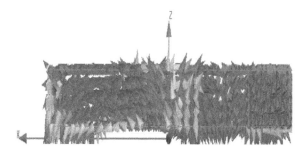

FIGURE A2.3   Resonant mode excited into DRA.

FIGURE A2.4   Resonant mode in DRA.

Annexure A3: Design of Horn DRA

Design specifications for resonant frequency 2,4 GHz (Figure A3.1):
Taking RT duroid as substrate, $\varepsilon_r = 2.2$

$$d = 0.8 \text{ mm}$$

Characteristic impedance $Z_o = 50 \, \Omega$

$$f = 2.4 \text{ GHz}$$

Let, $\dfrac{w}{d} > 2$; the ratio $\left(\dfrac{\text{width}}{\text{height}}\right)$ of microstrip

$$B = \frac{377\pi}{2z_o\sqrt{\varepsilon_r}} = \frac{377 \times 3.14}{2 \times 50\sqrt{2.2}} = 7.985$$

$$\frac{w}{d} = \Sigma \frac{se^A}{e^{2A} - 2} \times \text{ratio}\left(\frac{\omega}{d} < 2\right)$$

$$\text{Ldra} = 0.4 \quad \lambda_o = 50 \text{ mm}$$

$$\frac{2}{\pi}\left[B - 1 - \ln(2B - 1) + \frac{\varepsilon_r - 1}{2\,\varepsilon_r}\Sigma \ln(b - 1) + 0.39.0.61\right]$$

$$\frac{w}{d} = \Sigma \frac{se^A}{e^{2A} - 2}$$

$$\frac{2}{\pi}\left[B - 1 - \ln(2B - 1) + \frac{\varepsilon_r - 1}{2\,\varepsilon_r}\Sigma \ln(b - 1) + 0.39.0.61\frac{\omega}{d} > 2\right]$$

$$A = \frac{20}{60}\sqrt{\frac{\varepsilon_r + 1}{2}} + \frac{\varepsilon_r - 1}{\varepsilon_r + 1}\left(0.23 + \frac{0.11}{\varepsilon_r}\right): A \text{ and } B \text{ are constants}$$

When $\dfrac{w}{d} > 2$; microstrip width and height

$$\frac{w}{d} = \Sigma \frac{2}{3.14}[7.985 - 1 - \ln(2(7.985) - 1) + \frac{2.2 - 1}{2(2.2)}$$

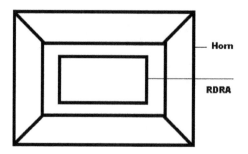

FIGURE A3.1 Horn DRA antenna.

$$\left\{ \ln\left(7.985-1\right)+0.39-\frac{0.61}{2.2}\right\}$$

$$w\left(\text{microstrip width}\right)=0.2\,\lambda_o=25\text{ mm}$$

$$\frac{w}{d}=0.391\text{ cm}=0.039\text{ mm}$$

$$w=0.03\times0.8\text{ mm}=2.4\text{ mm}=2.46\text{ mm}$$

Stub

$$Lst=\frac{\lambda_g}{8}\rightarrow\frac{51.68}{8}=6.45\text{ mm}\approx6.023\text{ mm}$$

$$\lambda_g=\frac{\lambda_o}{\sqrt{\varepsilon_{\text{reff}}}}$$

$$\varepsilon_{\text{reff}}=\frac{10+5.5+2.2}{3}=2.42$$

$$\lambda_g=\frac{12.5}{2.42}=51.65\text{ mm}$$

Slot
Stub length = $0.4\,\lambda_g=\frac{0.4\,\lambda_o}{\sqrt{\varepsilon_{\text{reff}}}}=0.4\times51.65-20.66\text{ mm}=20.24\text{ mm}$
Stub width $\rightarrow0.2\times20.24=4.04\text{ mm}$
Ground plane $\rightarrow$ 4 times the length of DRA $=4\times\left(50\text{ mm}\right)=200\text{ mm}$

$$\omega=200\text{ mm}$$

Inner DRA made up of Roger material $[\varepsilon_r = 10]$

$$L = 0.8\ \lambda_o = 100\ \text{mm}$$
$$\omega = 0.4\ \lambda_o = 50\ \text{mm}$$

Overall height = $0.338\ \lambda_o = 42.25$ mm

$$c = v\lambda\quad 3\times10^{11} = 2.4\times10^{9}\ \lambda_o$$

$$\lambda_o = \frac{3000}{24} = 12.5\ \text{mm}$$

Outer DRA is made of glass.

$$\left[\varepsilon_r = 5.5\right]$$

$$l = 0.496\ \lambda_o = 62\ \text{mm}$$
$$w = 0.28\ \lambda_o = 35\ \text{mm}$$

Upper rectangle dimension:

$$l2 = 0.88\ \lambda_o = 110\ \text{mm}$$
$$w2 = 0.98\ \lambda_o = 60\ \text{mm}$$

Annexure A4: Radiation Pattern of DRA

A4.1 RADIATION PATTERN

The radiation pattern of dielectric resonator antenna (DRA) depends upon the structure, shape, input excitation, point of excitation, boundary conditions, and material used. Frequency, magnitude, and phase of radiation are measured. Resonant modes can be controlled; hence, radiation pattern can also be controlled. Boundary conditions of DRA also affect radiation into DRA. Bore sight, broad side or end fire array patterns can be achieved in DRA. Directive or omnidirectional patterns can also be achieved based on structure and controlled excitation with controlled boundary conditions. Volume of cavity and shape of cavity also impact radiation patterns of DRA. The ground plane plays an important role in gaining total control of the radiation pattern of DRA. Also, the boundary conditions of DRA can further have control on radiation patterns along with mode generation. Circuit model values (capacitance, inductance, resistance and transformer ratios) of DRA can be derived to show the true behavior of the system as parallel R, L, C circuit. The meta surface-based DRA is developed to produce a collimated beam using the vortex concept. These 2D phase gradient meta surfaces can impart a lens effect for beam shaping. They work on the principle of OAM (orbital angular momentum). They can provide control on transmittance and transmission phase to provide desired collimation effects by using a plasmonic approach. For example, 8-element circular array of linear antennas provided a collimation beam effect, i.e., manipulation of the amplitude and phase of EM waves. In addition, DRA with concave shape, v shape, and convex shape can also be used for different types of beam formations with conventional materials. Also, an amalgamation of meta surface with concave and convex shaped DRAs are the best options for beam control (Figure A4.1).

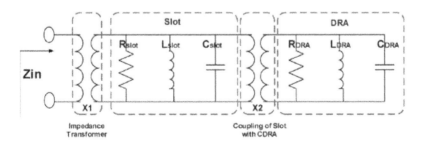

FIGURE A4.1 R, L, C equivalent circuit of DRA.

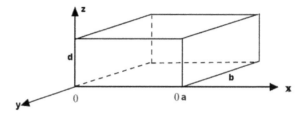

FIGURE A4.2   Rectangular DRA.

A4.2   RECTANGULAR WAVE GUIDE SOLUTION

Figure A4.2 has been shown as basic building block of rectangular DRA with a,b and d dimensions.

Helmholtz equation:

$$\nabla^2 \psi + k^2 \psi = 0 \text{ (source less medium)}$$

$$\nabla^2 \psi + k^2 \psi = -\mu \, j \text{ (medium with source)}$$

Maxwell's equations:

$$\nabla \times E = -\mu \frac{\partial H}{\partial t}$$

$$\nabla \times H = j + \frac{\partial E}{\partial t}$$

Solving L.H.S. on both sides first

$$\nabla \times E = \begin{vmatrix} i & j & k \\ \dfrac{\partial}{\partial x} & \dfrac{\partial}{\partial y} & \dfrac{\partial}{\partial z} \\ E_x & E_y & E_z \end{vmatrix} = i\left(\dfrac{\partial E_z}{\partial y} - \dfrac{\partial E_y}{\partial z} \right) - j\left(\dfrac{\partial E_z}{\partial x} - \dfrac{\partial E_x}{\partial z} \right) + k\left(\dfrac{\partial E_y}{\partial x} - \dfrac{\partial E_x}{\partial y} \right)$$

$$\nabla \times H = \begin{vmatrix} i & j & k \\ \dfrac{\partial}{\partial x} & \dfrac{\partial}{\partial y} & \dfrac{\partial}{\partial z} \\ H_x & H_y & H_z \end{vmatrix} = i\left(\dfrac{\partial H_z}{\partial y} - \dfrac{\partial H_y}{\partial z} \right) - j\left(\dfrac{\partial H_z}{\partial x} - \dfrac{\partial H_x}{\partial z} \right) + k\left(\dfrac{\partial H_y}{\partial x} - \dfrac{\partial H_x}{\partial y} \right)$$

Comparing with R.H.S. in both equations and getting value of H_x, H_y, H_z and E_x, E_y, E_z, we get:

$$Hx = \frac{1}{-j\omega\mu}\left(\frac{\partial E_z}{\partial y} - \frac{\partial E_y}{\partial z}\right) \qquad (A4.1)$$

$$Hy = \frac{1}{j\omega\mu}\left(\frac{\partial E_z}{\partial x} - \frac{\partial E_x}{\partial z}\right) \qquad (A4.2)$$

$$Hz = \frac{1}{-j\omega\mu}\left(\frac{\partial E_y}{\partial x} - \frac{\partial E_x}{\partial y}\right) \qquad (A4.3)$$

$$Ex = \frac{1}{j\omega\varepsilon}\left(\frac{\partial H_z}{\partial y} - \frac{\partial H_y}{\partial z}\right) \qquad (A4.4)$$

$$Ey = \frac{1}{-j\omega\varepsilon}\left(\frac{\partial H_z}{\partial x} - \frac{\partial H_x}{\partial z}\right) \qquad (A4.5)$$

$$Ez = \frac{1}{j\omega\varepsilon}\left(\frac{\partial H_y}{\partial x} - \frac{\partial H_x}{\partial y}\right) \qquad (A4.6)$$

Substituting $-\dfrac{\partial}{\partial z} = \gamma$;

$$Hx = \frac{j\omega\varepsilon\dfrac{\partial E_z}{\partial y} + \gamma\dfrac{\partial H_z}{\partial x}}{\gamma^2 + \omega^2\mu\varepsilon} \qquad Ex = \frac{-j\omega\mu\dfrac{\partial H_z}{\partial y} + \gamma\dfrac{\partial E_z}{\partial x}}{\gamma^2 + \omega^2\mu\varepsilon}$$

$$Hy = \frac{-j\omega\dfrac{\partial E_z}{\partial x} + \gamma\dfrac{\partial H_z}{\partial x}}{\gamma^2 + \omega^2\mu\varepsilon} \qquad Ey = \frac{j\omega\mu\dfrac{\partial H_z}{\partial x} + \gamma\dfrac{\partial E_z}{\partial y}}{\gamma^2 + \omega^2\mu\varepsilon}$$

$$-(\gamma^2 + \omega^2\mu\varepsilon)\,Hz = \frac{\delta^2 H_z}{\delta x^2} + \frac{\delta^2 H_z}{\delta y^2}$$

$$-(\gamma^2 + \omega^2\mu\varepsilon)\,Ez = \frac{\delta^2 E_z}{\delta x^2} + \frac{\delta^2 E_z}{\delta y^2}$$

On looking at the above equations, we get that H_z, E_z in a 2D Helmholtz equation. Now, rewriting the Helmholtz equation for source free medium:

$$\nabla^2\psi + k^2\psi = 0$$

Here, k is the wave number.

$$\Psi = X(x)Y(y)Z(z)$$

$$\left(\frac{1}{X}\left(\frac{d^2X}{dx^2}\right)+\frac{1}{Y}\left(\frac{d^2Y}{dy^2}\right)+\frac{1}{Z}\left(\frac{d^2Z}{dz^2}\right)\right)+k^2 = 0$$

Separating the independent terms, we get:

$$\frac{1}{X}\left(\frac{d^2X}{dx^2}\right) = -k_x^2$$

$$\frac{1}{Y}\left(\frac{d^2Y}{dy^2}\right) = -k_y^2$$

$$\frac{1}{Z}\left(\frac{d^2Z}{dz^2}\right) = -k_z^2$$

$$k^2 = k_x^2 + k_y^2 + k_z^2$$

$$\Psi = \left\{\left(A\sin k_x.x + B\cos k_x x\right)\left(C\sin k_y.y + D\cos k_y.y\right)\right\}e^{-jk_z z}$$

Solving the above function and keeping propagation in the $+z$ direction only, we get:
   TE mode:

$$Hz = \sum_{m,n}\left\{C_{mn}\left(\cos\frac{m\pi x}{a}\right)\left(\cos\frac{n\pi y}{b}\right)\right\}e^{-jk_z z}; \; C_{mn} \text{ Fourier Coefficients} \qquad \text{(A4.7)}$$

TM mode:

$$Ez = \sum_{m,n}\left\{D_{mn}\left(\sin\frac{m\pi x}{a}\right)\left(\sin\frac{n\pi y}{b}\right)\right\}e^{-jk_z z}; \; D_{mn} \text{ Fourier Coefficients} \qquad \text{(A4.8)}$$

These Fourier coefficients are resultant of mode amplitude and propagation constant at any instant.
   Let $\gamma = -jk_z$ and m, n be integers and a, b be dimensions;

$$\left(\frac{m\pi}{a}\right)^2 + \left(\frac{n\pi}{b}\right)^2 = (k_c)_{mn}; \text{ cut-off frequency}$$

$$k_z^2 = \omega^2\mu\epsilon - \left(\left(\frac{m\pi}{a}\right)^2 + \left(\frac{n\pi}{b}\right)^2\right)$$

Hence, EM waves will propagate in the z-direction if

$$\omega^2 \mu\epsilon - \left(\left(\frac{m\pi}{a}\right)^2 + \left(\frac{n\pi}{b}\right)^2\right) > 0$$

This gives cut-off frequency as

$$\omega_c = \frac{1}{\sqrt{\mu\epsilon}} \sqrt{\left(\left(\frac{m\pi}{a}\right)^2 + \left(\frac{n\pi}{b}\right)^2\right)}$$

This means the waveguide will support all waves having ω greater than ω_c to propagate.

Now, rewriting H_z and E_z

$$Hz = \sum_{m,n} \left\{ C_{mn} \left(\cos\frac{m\pi x}{a} \right) \left(\cos\frac{n\pi y}{b} \right) \right\} e^{-jk_z z}$$

$$Ez = \sum_{m,n} \left\{ D_{mn} \left(\sin\frac{m\pi x}{a} \right) \left(\sin\frac{n\pi y}{b} \right) \right\} e^{-jk_z z}$$

Here, C_{mn} & D_{mn} are coefficients of Hz and Ez fields.

$E_{ix(x,y)}$ Incident EM wave in x direction;
$E_{iy(x,y)}$ Incident EM wave in y direction;

$$E_{ix(x,y)} = \sum \left[\frac{j\omega\mu D_{(m,n)}\left(\frac{n\pi}{b}\right) + \gamma_{m,n} C_{(m,n)}\left(\frac{m\pi}{a}\right)}{h^2_{m,n}} \right] \cos\left(\frac{m\pi x}{a}\right) \sin\left(\frac{n\pi y}{b}\right) \exp\left(-\gamma_{m,n} z\right);$$

Similarly,

$$E_{iy(x,y)} = \sum \left[\frac{j\omega\mu D_{(m,n)}\left(\frac{m\pi}{a}\right) + \gamma_{m,n} C_{(m,n)}\left(\frac{n\pi}{b}\right)}{h^2_{m,n}} \right] \cos\left(\frac{m\pi x}{a}\right) \sin\left(\frac{n\pi y}{b}\right) \exp\left(-\gamma_{m,n} z\right);$$

On simplification,

$$E_{ix(m,n)} = \frac{j\omega\mu D_{(m,n)}\left(\dfrac{n\pi}{b}\right) + \gamma_{m,n} C_{(m,n)}\left(\dfrac{m\pi}{a}\right)}{h^2_{m,n}}$$

Similarly,

$$E_{iy(m,n)} = \frac{j\omega\mu D_{(m,n)}\left(\dfrac{m\pi}{a}\right) + \gamma_{m,n} C_{(m,n)}\left(\dfrac{n\pi}{b}\right)}{h^2_{m,n}}$$

$$\begin{bmatrix} E_{ix(m,n)} \\ E_{iy(m,n)} \end{bmatrix} = \begin{bmatrix} \dfrac{m\pi}{a}\dfrac{\gamma_{(m,n)}}{h^2_{m,n}} & \dfrac{n\pi}{b}\dfrac{j\omega\mu}{h^2_{m,n}} \\ -\dfrac{\gamma_{(m,n)}}{b}\dfrac{n\pi}{h^2_{m,n}} & \dfrac{-j\omega\mu}{a}\dfrac{m\pi}{h^2_{m,n}} \end{bmatrix} \begin{bmatrix} C_{mn} \\ D_{mn} \end{bmatrix}; \text{ we can now get the value of } C_{mn},$$

D_{mn} after substitution of $E_{ix(m,n)}$, $E_{iy(m,n)}$ values where $h^2_{m,n} = \left(\dfrac{m}{a}\right)^2 + \left(\dfrac{n}{b}\right)^2$ and

$$\gamma_{m,n} = \sqrt{h^2_{m,n} - \omega^2 \mu\varepsilon}.$$

Hence, $C_{(m,n)}$ & $D_{(m,n)}$ gives us relative amplitudes of Ez and Hz fields in TM or TE modes.

Therefore, we get the solution of possible amplitudes and phase of waves propagating through rectangular waveguides called modes of propagation.

The half-wave Fourier expansion in waveguide is given as follows:

$$f_{m,n} = \int_0^a \cos\left(\frac{m\pi x}{a}\right)\cos\left(\frac{m'\pi x}{a}\right)dx = \int_0^b \sin\left(\frac{n\pi y}{b}\right)\sin\left(\frac{n'\pi y}{b}\right)dy;$$

in even or odd terms, i.e.,
   $f(x) = f(-x)$ for even term (all cosine terms) or even modes
   where m, m' and n, $n' \geq 1$

$$E_{ix(m,n)} = \frac{2}{ab}\int_0^a\int_0^b E_{ix(x,y)}\left(\cos\frac{m\pi x}{a}\right)\left(\cos\frac{n\pi y}{b}\right)dxdy$$

$$E_{iy(m,n)} = \frac{2}{ab}\int_0^a\int_0^b E_{iy(x,y)}\left(\sin\frac{m\pi x}{a}\right)\left(\sin\frac{n\pi y}{b}\right)dxdy$$

Half-wave Fourier analysis will have odd or even terms, i.e., sine–sine or cosine–cosine.

If $f(x) = f(-x)$, even harmonics will take place and only cosine terms will occur, i.e.,

$$f(x) = \sum_{n=1}^{\infty} C_n \cos\left(\frac{\pi n x}{a}\right)$$

where $C_n = \dfrac{2}{a}\displaystyle\int_0^a f(x)\cos\left(\frac{n\pi}{a}x\right)dx$

Similarly, for odd terms, $f(x) \neq f(-x)$;

$$f(x) = \sum_{n=1}^{\infty} B_n \sin\left(\frac{\pi n x}{a}\right)$$

where $B_n = \dfrac{2}{a}\displaystyle\int_0^a f(x)\sin\left(\frac{n\pi}{a}x\right)dx$

A4.3　BEAM CONTROL

Beam width control of DRA can be obtained by using DRA of concave and convex nature attached at the top of rectangular DRA. **V- and U-shaped DRA**, if placed on top of rectangular DRA, shall provide narrow beam width with higher gain. Reverse of **V shape DRA** placed on top of rectangular DRA can provide more beam width radiation pattern, hence less gain. Rectangular DRA can also provide bore sight beams in radiation pattern.

Solving the wave equation with boundary conditions $E_{tan} = 0$, we find E-fields and then H-fields. Now that the shape and size of resonator is given, wave equation shall give the solution of characteristic frequencies ω (*mnp*) called eigen values or eigen frequencies of electromagnetic oscillations of a cavity resonator.

Lowest Eigen frequnecy ω_1 is $\frac{c}{l}$, where l is the dimension of resonator.
Higher frequency ($\omega \gg \frac{c}{l}$); then ω is $\frac{v\omega^2}{2\pi^2 c^3}$.
Hence, it depends on volume and net on shape of resonator.

For resonator: $\displaystyle\sum_{m,n,r} f_{m,n,r} \sin\left(\frac{\pi m x}{a}\right)\sin\left(\frac{\pi n y}{b}\right)\sin\left(\frac{\pi r z}{d}\right) = f(x, y, z)$

$\left(\dfrac{\partial^2}{\partial x^2} + \dfrac{\partial^2}{\partial y^2} + \dfrac{\partial^2}{\partial z^2}\right)\psi(x, y, z) + k^2\psi(x, y, z) = f(x, y, z)$; Helmholtz equation

$\psi(x, y, z) = \displaystyle\sum_{m,n,r} C_{mnr} \sin\left(\frac{\pi m x}{a}\right)\sin\left(\frac{\pi n y}{b}\right)\sin\left(\frac{\pi r z}{d}\right)$

$$\left[k^2 - \pi^2 \left(\frac{m^2}{a^2} + \frac{n^2}{b^2} + \frac{r^2}{d^2} \right) \right] C_{mnr} = f_{mnr}$$

Amplitude coefficient, $C_{mnr} = \dfrac{f_{mnr}}{k^2 - \pi^2 \left(\dfrac{m^2}{a^2} + \dfrac{n^2}{b^2} + \dfrac{r^2}{d^2} \right) \Big/ \omega_{mnr}}$

$$k = \omega \sqrt{\mu \epsilon}$$

$$\omega = \omega(mn\,p) + \delta;$$

where δ is small deviation and r is different from p.

Hence,
$$C_{mnr} = \frac{f_{mnr}}{(\omega(mn\,p) + \delta)^2 - \omega(mnr)^2}$$

$$= \frac{f_{mnr}}{\delta(\omega(mn\,p) - \omega(mnr))}$$

A4.4   SIMPLE WIRE AS AN ANTENNA (FIGURE A4.3)

$$x''(t) + \omega_0^2\, x(t) = B\, e^{j\omega t}$$

$$x(t) = A e^{j\omega t}$$

$$\left(\omega_0^2 - \omega^2 \right) A = B$$

Hence, $A = \dfrac{B}{\omega_0^2 - \omega^2}$

$x(t) = \dfrac{B\, e^{j\omega t}}{\omega_0^2 - \omega^2}$, if $\omega_0 = \omega$; then $x(t)$ will be ∞

Now, $\omega = \omega_0 + \delta$ when δ is small deviation.

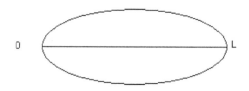

FIGURE A4.3   Simple radiating wire.

$$= \frac{B e^{j\omega t}}{(\omega_0 + \omega)(\omega_0 - \omega)}$$

$$= \frac{B e^{j\omega t}}{\delta(2\omega_0)}$$

$$\left(\frac{\partial^2}{\partial x^2} - \frac{1}{c^2}\frac{\partial^2}{\partial t^2}\right) u(x,t) = 0; \text{ at boundaries}$$

$$u(0,t) = 0 \text{ and } u(L,t) = 0$$

Taking the Fourier transform of the above equation,

$$\left(\frac{\partial^2}{\partial x^2} + \frac{\omega^2}{c^2}\right) \hat{u}(x,\omega) = 0$$

Writing the above terms in sine and cosine form, we have:

$$A \sin\left(\frac{\omega x}{c}\right) + B \cos\left(\frac{\omega x}{c}\right) = 0$$

$$\hat{u}(o,\omega) = 0$$

$$\hat{u}(L,\omega) = 0$$

$$\sin\left(\frac{\omega L}{c}\right) = 0; \text{ hence } k L = n\pi; \text{ sine values to be zero.}$$

$$\omega = kc = \frac{n\pi c}{L}, \text{ when } n = 1,2,3\ldots \text{ where } k = \omega/c.$$

When $2L$, it is fundamental frequency ω_1.
When L, the frequency is $2\omega_1$.
When $2L/3$, the frequency $3\omega_1$;

which can be generalized as:

$$\sum_n C(n) \sin\left(\frac{n\pi x}{L}\right)$$

A4.5   SIMPLE RADIATING SURFACE

$$\frac{\partial^2 \psi (x,y,t)}{\partial x^2} + \frac{\partial^2 \psi (x,y,t)}{\partial y^2} - \frac{1}{c^2}\frac{\partial^2 \psi (x,y,t)}{\partial t^2} = 0$$

Applying boundary conditions (Figure A4.4),

$$\psi (o, y,t) = \psi (a, y,t) = 0$$

$$\psi (x,0,t) = \psi (x,b,t) = 0$$

Let input excitation be some tension T,

$$\sigma dx dy \frac{\partial^2 \psi}{\partial t^2} = \frac{\partial}{\partial x}\left(T.dy \frac{\partial \psi}{\partial x} \right)dx + \frac{\partial}{\partial y}\left(Tdx \frac{\partial \psi}{\partial y} \right)dy$$

$$\frac{Y''}{Y} = -k_Y^2 ; \quad \frac{X''}{X} = -k_X^2 ;$$

Now from the Helmholtz equation (Figure A4.5):

$$\frac{\partial^2 \psi}{\partial t^2} - c^2 \nabla^2 \psi = 0$$

Using separation of variables:

$$\psi (x,y,t) = X(x)Y(y)T(t)$$

$$-\omega^2 = \frac{T''(t)}{T(t)} = c^2 \left(\frac{X''(x)}{X(x)} + \frac{Y''(y)}{Y(y)} \right) \qquad (A4.9)$$

y=b

x=a

FIGURE A4.4   Radiating surface.

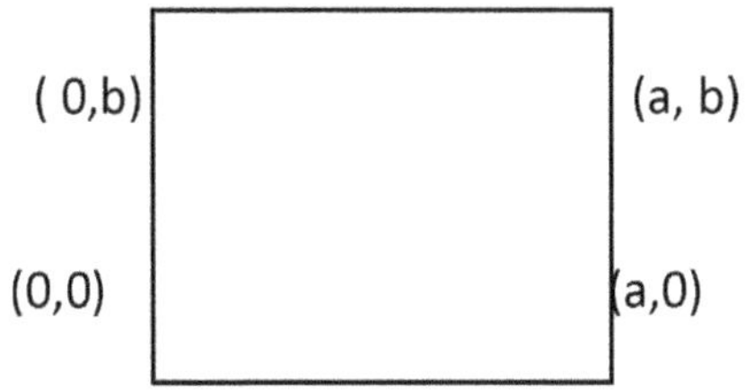

FIGURE A4.5 Resonator surface coordinates.

let: $X(x) = \sin(k_x x)$

$$Y(y) = \sin(k_y y)$$

$$k_x^2 + k_y^2 = \frac{\omega^2}{c^2}$$

where k_x and k_y can be

$$k_x = \frac{m\pi}{a}; \, k_y = \frac{n\pi}{b}$$

Equation (A4.1) can be written as $\omega(mn) = c\pi \sqrt{\left(\frac{m}{(a)}\right)^2 + \left(\frac{n}{b}\right)^2}$.

From Fourier series analysis,

$$\omega(m,n) = \sum_{m,n=1}^{\infty} \sin\left(\frac{m\pi x}{a}\right) \sin\left(\frac{n\pi y}{b}\right) \left[C(m,n) e^{j\omega(m,n)t} + D(m,n) e^{-j\omega(m,n)t} \right]$$

(A4.10)

At $t = 0$; $\psi(x,y,o) = \psi_o(x,y)$

On differentiating equation $\psi_o(x,y)$ we get, $\psi_1(x,y,o) = \psi_1(x,y)$

When $t \neq 0$;

$$\psi_0(x,y) = \sum_{m,n=1}^{\infty} (C(m,n) + D(m,n)) \sin\left(\frac{m\pi x}{a}\right) \sin\left(\frac{n\pi y}{b}\right) \quad \text{(A4.11)}$$

$$\psi_1(x,y) = \sum_{m,\,n=1}^{\infty} j\omega(m,n)(C(m,n) - D(m,n)) \sin\left(\frac{m\pi x}{a}\right) \sin\left(\frac{n\pi y}{b}\right) \quad \text{(A4.12)}$$

$$\frac{2}{\sqrt{ab}} \int_0^a \int_0^b \psi_0(x,y) \sin\left(\frac{m\pi x}{a}\right) \sin\left(\frac{n\pi y}{b}\right) dx\, dy = \left[C(m,n) + D(m,n) \right] \quad \text{(A4.13)}$$

Similarly

$$j\omega(m,n)\frac{2}{\sqrt{ab}}\int\psi_1(x,y)\sin\left(\frac{m\pi x}{a}\right)\sin\left(\frac{n\pi y}{b}\right)dx\,dy=\left[C(m,n)-D(m,n)\right] \quad (A4.14)$$

Hence, we obtain the value of $C\ (m,n),D\ (m,n)$ from equations (A4.3) and (A4.4).

$$\left[C(m,n),D(m,n)\right]=$$

$$\frac{1}{\sqrt{ab}}\left[\iint\psi_0(x,y)\sin\left(\frac{m\pi x}{a}\right)\sin\left(\frac{n\pi y}{b}\right)dx\,dy\pm\frac{1}{j\omega(m,n)}\int\psi_*(x,y)\sin\left(\frac{m\pi x}{a}\right)\right.$$
$$\left.\sin\left(\frac{n\pi y}{b}\right)dx\,dy\right]$$

$$(A4.15)$$

Hence, from equation (A4.15),

$$\psi_0(x,y)=A\sin\left(\frac{m_0\pi x}{a}\right)\sin\left(\frac{n_0\pi y}{b}\right)$$

$$\psi_1(x,y)=B\sin\left(\frac{m_o\pi x}{a}\right)\sin\left(\frac{n_o\pi y}{b}\right)$$

Due to force, perturbation occurs (Figure A4.6).
   Solving equation (A4.15), we get:

$$\left(C(m,n),D(m,n)\right)=\delta\left[m-m_o\right]\delta\left[n-n_o\right]$$

$$=\left(\frac{A}{\sqrt{ab}}\left(\frac{a}{2}\times\frac{b}{2}\right)+\left(\frac{1}{\sqrt{ab}}\frac{B}{j\omega(m_o,n_o)}\right)\left(\frac{a}{2}\times\frac{b}{2}\right)\right.$$

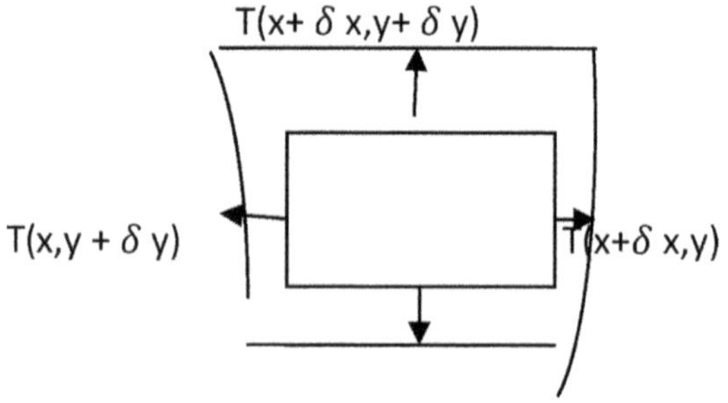

FIGURE A4.6   Deformation due to excitation.

$$\left(C(m,n),D(m,n)\right)=\sqrt{ab}\left(\frac{A}{4}\mp\frac{j.B}{4}\right)\frac{\sqrt{ab}}{4}\left(A\mp jB\right)\delta\left[m-m_0\right]\delta\left[n-n_o\right]$$

$$\psi(x,y,t)=\frac{\sqrt{ab}}{2}Re(A-jB)\sin\left(\frac{m_o\pi x}{a}\right)\sin\left(\frac{n_o\pi y}{b}\right)e^{j\omega(m_o,n_o)t}$$

Hence, we complete the solution of two-dimensional resonator.

$$\psi(x,y,t)=\frac{\sqrt{ab}}{2}\left(A\,\cos\left(\omega(m_0\,n_0)t\right)+B\,\sin\left(\omega(m_0\,n_0)t\right)\right)\sin\left(\frac{m_0\pi x}{a}\right)\sin\left(\frac{n_0\pi y}{b}\right)$$

$$(A4.16)$$

Alternate method:

$$m=2b\sin\theta;$$

$$n=2a\cos\theta;$$

Dividing both sides of the above equations by $2a$ and $2b$ and adding them gives us,

$$\frac{1}{\lambda^2}=\frac{n^2}{4a^2}+\frac{m^2}{4b^2};\text{ where }k^2=k^2_x+k^2_y$$

Thus, resonant frequency of resonator can be determined.
    Half-wave Fourier analysis:

$$f(x)=\frac{a_0}{2}+\sum_{n=1}^{\infty}\left[a_n\cos\left(\frac{2n\pi}{a}x\right)+b_n\sin\left(\frac{2n\pi}{a}x\right)\right]$$

$$a_n=\frac{2}{a}\int_0^a f(x)\cos\left(\frac{2n\pi}{a}x\right)dx$$

$$b_n=\frac{2}{a}\int_0^a f(x)\sin\left(\frac{2n\pi}{a}x\right)dx$$

Half-wave Fourier analysis will have odd or even terms, i.e., sine–sine or cosine–cosine. If $f(x)=f(-x)$, even harmonics will take place and only cosine terms will occur, i.e.,

$$fx=\sum_{n=1}^{\infty}C_n\cos\left(\frac{\pi nx}{a}\right)$$

where $C_n=\frac{2}{a}\int_0^a f(x)\cos\left(\frac{n\pi}{a}x\right)dx$

Similarly, for odd terms, $f(x) \neq f(-x)$;

$$fx = \sum_{n=1}^{\infty} B_n \sin\left(\frac{\pi nx}{a}\right)$$

where $B_n = \dfrac{2}{a} \displaystyle\int_0^a f(x)\sin\left(\frac{n\pi}{a}x\right)dx$

A4.6   FREQUENCY SPECTRAL OF EM WAVES

Every wave can be subjected to the process of spectral resolution, i.e., it can be represented as a superposition of monochromatic waves of various frequencies.

This is the usual expansion in Fourier series. It contains the frequencies which are integral multiples of the "fundamental" frequency $\omega_o = \dfrac{2\pi}{T}$, where T is the period of the field. We therefore write it in the following form.

$$f = \sum_{n=-\infty}^{\infty} f_n e^{-j\omega_o nt}$$

where f is any of the quantities describing the field. The quantities f_n are defined in terms of the function f by the integrals.

$$f_n = \frac{1}{T} \int_{-T/2}^{T/2} f(t) e^{jn\omega_o t} \; dt.$$

Because $f(t)$ must be real,

$$f_n = f_n^*.$$

Let us express the total intensity of the wave, i.e., the integrals of f^2 over all time in terms of the intensity of the Fourier components. Now, we have

$$\int_{-\infty}^{\infty} f^2 dt = \int_{-\infty}^{\infty} \left\{ f \int_{-\infty}^{\infty} f_\omega e^{-i\omega t}\frac{d\omega}{2\pi} \right\} dt = \int_{-\infty}^{\infty} \left\{ f_\omega \int_{-\infty}^{\infty} f_e^{-i\omega t}\, dt \right\}\frac{d\omega}{2\pi} = \int_{-\infty}^{\infty} f_\omega f_{-\omega}\frac{d\omega}{2\pi},$$

$$\int_{-\infty}^{\infty} f^2 dt = \int_{-\infty}^{\infty} |f_\omega|^2 \frac{d\omega}{2\pi} = 2\int_0^{\infty} |f_\omega|^2 \frac{d\omega}{2\pi}.$$

$f(t) = \frac{1}{2\pi}\int_{-\infty}^{\infty} f_\omega e^{-i\omega t}\, d\omega$, where the Fourier components are given in terms of the function $f(t)$ by the integrals, $f_\omega = \int_{-\infty}^{\infty} f(t)e^{i\omega t}dt$.

A4.7 POWER AND ENERGY SIGNALS

Let $x(t)$ be the input signal, i.e., voltage signal. As per Parseval's power theorem, energy associated with this signal will be

$$E = \int_{-\infty}^{\infty} |x(t)|^2 \, dt; \text{ in time domain}$$

$$= \frac{1}{2\pi} \int_{-\infty}^{\infty} |X(\omega)|^2 \, d\omega; \text{ in frequency domain}$$

The amount of energy radiated by this signal, when applied across antenna having radiation resistance Rr, shall be (Figure A4.7)

$$E = \frac{1}{Rr} \int_{-\infty}^{\infty} |x(t)|^2 \, dt = \frac{1}{2\pi Rr} \int_{-\infty}^{\infty} |X(\omega)|^2 \, d\omega$$

Now if the input signal is $x(t)$ having current signal

$$E = R_r \int_{-\infty}^{\infty} |(x(t)|^2 \, dt = \frac{R}{2\pi} \int_{-\infty}^{\infty} |X(\omega)|^2 \, d\omega$$

ESD energy spectral density; energy spread per unit volume across 1 Ω resistor,

$$\text{ESD} = |X(\omega)|^2$$

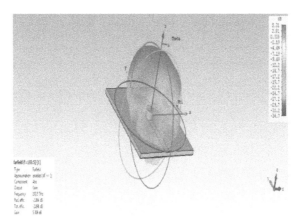

FIGURE A4.7 Radiation pattern of DRA.

DFT discrete Fourier transform in time domain into frequency domain spectral analysis,

$$X(k) = \sum_{n=0}^{N-1} x(n)e^{\frac{-j2\pi nk}{N}}; \quad k = 0, 1 \& N - 1.$$

$$X(n) = \frac{1}{N}\sum_{k=0}^{N-1} X(k) \, e^{\frac{-j2\pi kn}{N}}; \quad n = 0, 1, 2, \ldots N - 1.$$

$X(n)$ finite sequence.

DFT has finite length N, period N,

$$\psi(\theta, \phi) = k\left(\vec{r}_n . \vec{r}_0\right) = (n-1)kd\sin\theta$$

$$E(\sin\theta) = \sum_{n=1}^{N} e^{j(n-1)(kd\sin\theta)}$$

Annexure A5: DRA Modes, Bandwidth and Beam Width Control

The bandwidth of DRA can be controlled by merging of adjacent modes. Low permittivity substrates are also used for wide band response in DRA. DRA stacking-based structures are also used to achieve wide band response. Turtle shaped DRA has wide band response. Concave- and convex- shaped DRA can provide beam width control as well as bandwidth control at operating frequency. The concave shape can be tilted to the right and left to control the beam in specific directions.

The band-notch is obtained by applying the graphene nano-ribbons at the top of the radiating metallic patch or DRA. The notched frequency band can be tuned by changing the chemical potential of graphene nano-ribbons, keeping the cut-off frequencies of antenna response unchanged. Multiple resonant modes thus can be merged or separated for bandwidth and multimode requirements. Addition of graphene nano-ribbons in the antenna structure provides the dip of real part of impedance in the notched band region. The addition of graphene ribbons shifts the resonant frequency of these modes to the lower side. The best technique to control beam width of DRA is use of low permittivity material layers at the slot side and higher permittivity layers toward the top of DRA in a progressive manner. This will make e.m. fields to bend toward normal, hence resulting in beam width narrowing, and the reverse can be used for beam width widening. A second ground plane can also be used to get further narrow beam width. Metal plates in vertical format connected with ground plane can also be used to narrow down the beam width. Use of superstrate of horn DRA will make this beam width further narrower. Use of higher order modes in DRA for achieving narrow beam is also a good technique. Hence, wide beam width as well as narrow beam width can be achieved with proper design of DRA.

Smart dielectric resonator antenna (DRA) having beam control mechanism is a new area to be explored by antenna researchers. RDRA has low loss, design flexibility, high efficiency, compact size, and desired radiated beam control. Developing beam control in RDRA has been investigated for the first time in this letter. Narrow beam width is a requirement for line of sight communications (LOS), satellite communication, radar and wireless local area network (WLAN). A unique technique for beam control and beam width control has been proposed using pit top and mount top RDRA beam controlled from $\pm 30°$ to $\pm 70°$ in a broadside radiation pattern. Conformal RDRA designs have also been investigated. Hardware of U shape pit top RDRA, mount shape DRA, left side arc RDRA, right side arc shape RDRA, right

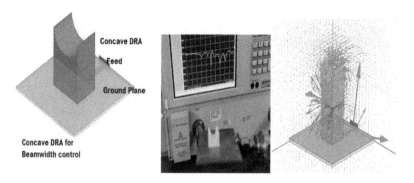

FIGURE A5.1   DRA concave shape with high permittivity on top material (narrow beam width, high directivity).

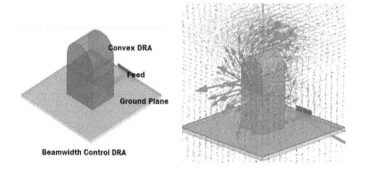

FIGURE A5.2   DRA convex shape with lower permittivity at bottom and high permittivity on top (narrow beam width, less directivity).

FIGURE A5.3   Rectangular DRA under test.

side mount RDRA and left side mount RDRA can be developed toward beam control and radiation pattern manipulation (Figures A5.1–A5.3). Hemispherical DRA can be used for large beam width. Conical nano DRA as shown in Figure A5.4 can provide narrow beam width stacks of low and high permittivity materials. Satellite, radar, and microwave communication need narrow beam width. Mobile and cellular

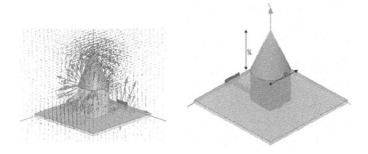

FIGURE A5.4   Cone DRA E-fields.

communication need wide beam widths. Narrowing of bandwidth at higher order modes has also been achieved.

Use of dielectric constants to bend e.m. wave toward normal has been applied in a progressive manner to achieve narrow beam width of DRA, a unique technique to control antenna beam width.

LOS communication and radar systems need narrow beam width because of high gain requirements. Nano antenna can provide narrow beam width as compared to microwave antennas. Phase shift arrays or MIMO antennas are generally used for narrow beam width requirements. Beam width control is an important phenomenon in antenna theory. Solid angle can be increased as well as decreased. Higher order modes have less beam width as compared to fundamental mode. Control of resonant modes has now become possible in DRA, hence beam width can be controlled in similar manner. Excitation of higher order mode and operating antenna at higher order mode is an excellent technique to narrow down the antenna beam width. This way passive designing of DRA shall be very helpful in controlling antenna beam width.

A5.1   DRA FORMULATIONS

$$\text{DR dimensions} = 24 \times 12 \times 24 \text{ mm}$$

$$\text{Slot dimensions} = 13 \times 1.5 \text{ mm}$$

$$\text{Substrate dimensions} = 80 \times 80 \times 0.8 \text{ mm (four times of DRA length)}$$

$$\text{Substrate permittivity} \left(\epsilon_s \right) = 4.4$$

$$\text{DR permittivity} \left(\epsilon_r \right) = 34$$

Take the average of permittivity of both.

Note: TMM, sapphire, and zircar (Al_2O_3) dielectric substrates (permittivity of 12.8, 10, 10) are also available in the open market.

DRA resonant frequency (rectangular):

$$(f)_r m,n,p = \frac{c}{2\pi\sqrt{\mu\epsilon}}\sqrt{\left[\left(\frac{m\pi}{a}\right)^2 + \left(\frac{n\pi}{b}\right)^2 + \left(\frac{p\pi}{d}\right)^2\right]}$$

Slot dimensions:

$$l_s = 0.4\lambda_o / \sqrt{\epsilon_{eff}}$$

where $\epsilon_{eff} = \dfrac{\epsilon_r + \epsilon_s}{2}$

$$w_s = 0.2l_s$$

The characterless equation of DRA is given as:

$$\varepsilon_r k_o^2 = k_x^2 + k_y^2 + k_z^2;$$

where k_o is the free space wave and k_x, k_y, k_z are the propagation constants in X, Y, Z direction, respectively. Also, $k_o^2 = \omega_o^2 \mu_o \varepsilon_o$, hence resonant frequency in free space can be determined based on free space wave number. To determine propagation constants, i.e., k_x, k_y, and k_z, knowledge of transcendental equation is required. The transcended equation is developed for RDRA when fields are propagating in the z-direction and given below as:

$$k_z \tan\left(k_z \frac{d}{2}\right) = \sqrt{(\varepsilon_r - 1)k_o^2 - k_z^2};$$

Electrical walls of DRA:

$$E_{tan} = n \times E = 0;$$

$$H_{nor} = n \cdot H = 0;$$

Magnetic walls of DRA:

$$H_{tan} = n \times H = 0;$$

$$E_{nor} = n \cdot E = 0;$$

The radiation due to short magnetic dipoles:

$$d = \sum e \cdot r$$

where d is the dipole moment, e is the charge, r is the distance between two charges

$$d = \frac{d}{dt}\sum e \cdot r = \sum e \cdot v$$

$$d'' = \frac{d}{dt}\sum e \cdot v$$

DRA bandwidth control can be made with generation of higher order modes and merging them to get wide bandwidth.

A5.2 DRA A, B, D DIMENSIONS, EXCITED WITH PROBE FEED

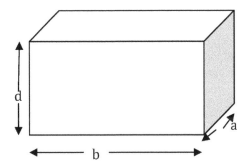

The rectangular DRA as cavity has dimensions as a, b, and d. Side walls have been taken as PMC magnetic conductors and top and bottom surfaces have been perfect electric conductors; theoretical fields (modes) solution have been obtained under boundary conditions with square type of feed probe for excitation.

E_X, $E_y = 0$, top and bottom plane being electric walls.
$H_X = 0$, side walls being magnetic walls.

$$H_z(x,y,z,t) = \sum_{mnp} \varepsilon(m,n,p)\frac{2}{\sqrt{ab}}\sin\left(\frac{m\pi x}{a}\right)\sin\left(\frac{n\pi y}{b}\right)\sin\left(\frac{p\pi z}{d}\right)$$

$$\cos\left(\omega(m,np)t + \phi(m,np)\right)$$

where m, n, p are eigen values (half wave variations in particular direction, i.e., x, y, z directions, respectively, a, b, d are the dimensions (width, length, height) of the cavity, $C(m,n,p)$ and $\phi(m,np)$ magnitude and phase coefficients of H_z and $D(m,n,p)$ and $\psi(m,np)$ for Ez.

Let $\dfrac{2}{\sqrt{ab}}\sin\left(\dfrac{m\pi x}{a}\right)\sin\left(\dfrac{n\pi y}{b}\right) = u_{mn}(x,y)$ for convenience.

$$E_z(x,y,z,t)=\sum_{mnp} D(m,n,p)\,\frac{2}{\sqrt{ab}}\,\cos\left(\frac{m\pi x}{a}\right)\cos\left(\frac{n\pi y}{b}\right)\sin\left(\frac{p\pi z}{d}\right)$$

$$\cos\left(\omega(m,np)t+\psi\,(m,np)\right)$$

Hence, a finite size antenna with a given type of excitation is a complicated boundary value problem which can be solved.

$$\underline{E}(x,y,z)=\underline{E}(X,Y)\exp(-\gamma Z)$$

$$\underline{H}(x,y,z)=\underline{H}(X,Y)\exp(-\gamma Z)$$

$$\underline{\nabla}\times\underline{E}=-j\omega\mu\,\underline{H}$$

Assuming sinusoidal dependence of E- and H-fields, we can resolve above fields in terms of transverse and longitudinal components.

Deriving transverse components only for $\underline{H}_\perp$

$$\left(\underline{\nabla}_\perp+\hat{z}\,\frac{\partial}{\partial z}\right)\times\left(\underline{E}_\perp+E_z\,\hat{z}\right)$$

$$=-j\omega\mu\left(\underline{H}_\perp+H_z\,\hat{z}\right)$$

$$\Rightarrow\underline{\nabla}_\perp\times\underline{E}_\perp=-j\omega\mu\,H_z\,\hat{z},$$

$$\underline{\nabla}_\perp\times\underline{E}_z\times\hat{z}+\hat{z}\times\frac{\partial E_\perp}{\partial z}$$

$$=-j\omega\mu\underline{H} \tag{A5.1}$$

$$\Rightarrow\underline{\nabla}_\perp\times\underline{E}_z\times\hat{z}\quad\gamma\hat{z}\times\underline{E}_\perp$$

$$=-j\omega\mu\,\underline{H}_\perp \tag{A5.2}$$

Likewise for $\underline{E}_\perp$

$$\underline{\nabla}\times\underline{H}=j\omega\epsilon\,\underline{E}$$

$$\Rightarrow\underline{\nabla}_\perp\times\underline{H}_\perp=j\omega\epsilon\,E\hat{z}, \tag{A5.3}$$

$$\underline{\nabla}_\perp\,H_z\times\hat{z}-\gamma\hat{z}\times\underline{H}_\perp=j\omega\epsilon\,\underline{E}_\perp \tag{A5.4}$$

In above calculations, we have derived transverse components of E and H in terms of

$$\underline{E}_\perp, \ \underline{H}_\perp, H_z, E_z.$$

Now, substituting for $\underline{H}_\perp$ from (A5.2) into (A5.4) gives

$$\underline{\nabla}_\perp H_z \times \hat{z} + \frac{\gamma \hat{z}}{jw\mu} \times \left(\underline{\nabla}_\perp E_z \times \hat{z} - \gamma \hat{z} \times \underline{E}_\perp \right) = jw \ \underline{E}_\perp \qquad (A5.5)$$

Or

$$\underline{\nabla}_\perp \ H_z \times \hat{z} + \frac{\gamma}{jw\mu} \underline{\nabla}_\perp \ E_z + \frac{\gamma^2}{jw\mu} \underline{E}_\perp = j \ w\underline{E}_\perp$$

Or

$$\Rightarrow \frac{h^2}{jw\mu} \ \underline{E}_\perp = -\left\{ \underline{\nabla}_\perp H_z \times \hat{z} + \frac{\gamma}{jw\mu} \underline{\nabla}_\perp \ E_z \right\}$$

Or

$$\underline{E}_\perp = -\frac{jw\mu}{h^2} \left\{ \underline{\nabla}_\perp H_z \times \hat{z} + \frac{\gamma}{i\omega\mu} \underline{\nabla}_\perp \ E_z \right\}$$

$$= \frac{\gamma}{h^2} \ \underline{\nabla}_\perp \ E_z - \frac{j\omega\mu}{h^2} \ \underline{\nabla}_\perp H_z \times \hat{z}$$

From equations (A5.2) and (A5.5),

$$-j\omega\mu\underline{H}_\perp = \underline{\nabla}_\perp \ E_z \times \hat{z} + \gamma \hat{z} \times \left\{ \frac{\gamma}{h^2} \ \underline{\nabla}_\perp E_z + \frac{j\omega\mu}{h^2} \ \underline{\nabla}_\perp \ H_z \times \hat{z} \right\}$$

$$= \left(\frac{\gamma^2}{h^2} -1 \right) \left(\hat{z} \times \underline{\nabla}_\perp \ E_z \right) \frac{j\omega\mu\gamma}{h^2} \underline{\nabla}_\perp \ H_z$$

$$= \frac{\omega^2 \beta \ \epsilon}{h^2} \left(\hat{z} \times \underline{\nabla}_\perp \ E_z \right) + \frac{j\omega\mu\gamma}{h^2} \underline{\nabla}_\perp \ H_z$$

Or

$$\underline{H}_\perp = -\frac{j\omega \ \epsilon}{h^2} \ (\hat{z} \times \underline{\nabla}_\perp \ E_z) - \frac{\gamma}{h^2} \underline{\nabla}_\perp H_z$$

Now, we specialize to a cylindrical wave guide of radius R;

$$\underline{E}_\perp = E_\rho \left(\rho, \phi \right) \hat{\rho} + \underline{E}_\phi \ \left(\rho, \phi \right) \hat{\phi},$$

$$H_{\perp} = H_{\rho}\left((\rho,\phi)\,\hat{\rho} + H_{\phi}\left(\rho,\phi\right)\hat{\phi},\right.$$

$$\nabla_{\perp} = \hat{\rho}\,\frac{\partial}{\partial\rho} + \frac{\hat{\phi}}{\rho}\,\frac{\partial}{\partial\phi}$$

$$\Rightarrow E_{\perp} = -\frac{\gamma}{h^2}\left(\hat{\rho}\,\frac{\partial}{\partial\rho} + \frac{\hat{\phi}}{\rho}\,\frac{\partial}{\partial\phi}\right)E_Z(\rho,\phi) - \frac{j\omega\mu}{h^2}\left(\hat{\rho}\,H_{z,\rho} + \frac{\hat{\phi}}{\rho}\,H_{z,\phi}\right)\times\hat{Z}$$

$$= \left(-\frac{\gamma}{h^2}\,E_{z,\rho}\,\frac{j\omega\mu}{h^2\rho}\,H_{z,\phi}\right)\hat{\rho} + \left(\left(-\frac{\gamma}{\rho h^2}\,E_{z,\phi} + \frac{j\omega\mu}{h^2}\,H_{z,\rho}\right)\hat{\phi}\right)$$

Thus

$$E_{\rho} = -\frac{\gamma}{h^2}\,E_{z,\rho} - \frac{j\omega\mu}{\rho h^2}\,H_{z,\phi}\Bigg) \qquad\qquad\text{(A5.6)}$$

$$E_{\phi} = -\frac{\gamma}{\rho h^2}\,E_{z,\phi} + \frac{jw\mu}{h^2}\,H_{z,\rho}$$

Likewise

$$H_{\perp} = \frac{-jw\epsilon}{h^2}\,\hat{Z}\times\left\{E_{z,\rho}\,\hat{\rho} + \frac{1}{\rho}E_{z,\phi}\hat{\phi}\right\} - \frac{\gamma}{h^2}\left\{H_{z,\rho}\,\hat{\rho} + \frac{1}{\rho}H_{z,\phi}\hat{\phi}\right\}$$

$$= \hat{\rho}\left\{\frac{j\omega\epsilon}{\rho h^2}\,E_{z,\phi} - \frac{\gamma}{h^2}\,H_{z,\rho}\right) + \hat{\phi}\left\{-\frac{jw\epsilon}{h^2}\,E_{z,\rho} - \frac{\gamma}{\rho h^2}\,H_{z,\phi}\right\}$$

The equation

$$\nabla_{\perp}\times E_{\perp} = -j\omega\mu\,H_z\,\hat{z},$$

$$\nabla_{\perp}\times H_{\perp} = j\omega\epsilon\,E_z\,\hat{z},$$

On substituting for $E_{\perp}$ and $H_{\perp}$ from (A5.5), we have

$$\left(\nabla_{\perp}^{\,2} + h^2\right)E_z = 0 \qquad\qquad\text{(A5.7a)}$$

$$\left(\nabla_{\perp}^{\,2} + h^2\right)H_z = 0 \qquad\qquad\text{(A5.7b)}$$

$$h^2 = \gamma^2 + \omega^2\varepsilon\mu$$

(A5.7a) and (A5.7b) can be expressed as

$$\frac{1}{\rho}\frac{\partial}{\partial\rho}\left(\rho E z_{\rho}\right)+\frac{1}{\rho^2}E_{z,\phi\phi}+h^2 E_z = 0$$

$$\frac{1}{\rho}\frac{\partial}{\partial\rho}\left(\rho H z_{\rho}\right)+\frac{1}{\rho^2}H_{z,\phi\phi}+h^2 H_z = 0$$

or equivalently

$$E_{z,\rho\rho}+\frac{1}{\rho}E_{z,\rho}+\frac{1}{\rho^2}E_{z,\phi\phi}+h^2 E_z = 0 \qquad (A5.8)$$

And likewise, for H_z.

Applying boundary conditions, $E_z\left(\rho,\Phi.\right)=0$ for $\rho = R$ and $H_\rho\left(\rho,\Phi.\right)=0$ for $\rho = R$

Let $E_z\left(\rho,\Phi.\right)= R\left(\rho\right)\Phi\left(\Phi\right)$ then from (A5.8)

$$\frac{R''(\rho)}{R(\rho)}+\frac{R'(\rho)}{\rho R(\rho)}+\frac{\Phi''(\Phi)}{\rho^2\Phi(\Phi)}+h^2 = 0$$

Or

$$\frac{\rho^2 R''(\rho)}{R(\rho)}+\frac{\rho R'(\rho)}{R(\rho)}+h^2\rho^2$$

$$=\frac{-\Phi''(\Phi)}{\Phi(\Phi)}=m^2$$

Here $\frac{-\Phi''(\Phi)}{\Phi(\Phi)}$ is independent of ρ^2, hence m^2 can be treated as constant.

Solving for $\frac{-\Phi''(\Phi)}{\Phi(\Phi)}$

Now $\Phi(\Phi)=A\cos(m\Phi)+ B\sin\,(n\Phi)$

$M\,\epsilon\,z_+$ for Φ to be single valued, hence for E_Z to be single valued

$$\rho^2 R''(\rho)+\left(h^2\rho^2 - m^2\right)R(\rho)=0$$

$$\Rightarrow R(\rho)=J_M\left(h\rho\right).$$

where J_M = Bessel functions of order m.

Since $E_z\left(R,\Phi\right)=0$, so we have

$$J_m\left(h,R\right)=0\Rightarrow h = h\,m, n =\frac{\alpha\,mn}{R}$$

$N \geq 1$ where $\{\alpha_{nm}\} n > 1$ are the roots of $J_m(x) = 0$

$$E_z(\rho, \Phi, z) = \exp(-h\, m, n^z)$$

$$\sum_{mn=1} J_m\left(\frac{\alpha mn}{R}\rho\right)(A[m,n]\cos(m\Phi) + B[m,n]\sin n\varnothing$$

$E_z(\rho, \Phi, z)$

$$= \sum_{mn} J_m\left(\frac{\alpha mn}{R}\rho\right)(A[m, n]\cos(m\Phi) + B[m,n]\sin(m\Phi)\exp(-Y[m, n]z)$$

where $\gamma[m,n] = \sqrt{h_{mn}^2 - w^2\mu \in} = \left(\frac{\alpha_{mn}^2}{R} - \omega^2\mu \in\right)$
Likewise, separation of variables gives

$$H_z(\rho, \Phi) = J_m(h, \rho)\left[\left(A\cos(m\Phi) + B\sin(m\Phi)\right)\right]$$

$$H_\rho(\rho, \Phi) = \frac{jw\epsilon}{\rho h^2}\ E_{z,\Phi}(\rho, \Phi) - \frac{\gamma}{h^2}H_z(\rho, \varnothing)$$

Now,

$$E_z(R, \varnothing) = 0 \Rightarrow E_{z,\varnothing}(R, \varnothing) = 0$$

Hence

$$H_z(R, \varnothing) = 0 \Rightarrow H_{z,\varnothing}(R, \varnothing) = 0$$

$$\Rightarrow J'_m(h, \rho)_{\rho=R} = J'_m(h, R) = 0$$

$$\Rightarrow h = \widetilde{h_{mn}}(tilda) = \frac{\beta_{mn}}{R}$$

where $\{\beta_{mn}\}$ $(n > 1)$ are the roots of $J'_m(x) = 0$. The complete solution for Hz is

$$H_z(\rho, \Phi, z) = \sum_{mn} J_m\left(\frac{\beta_{mn}\rho}{R}\right)(Cmn\ \cos m\varnothing + Dmn\ \sin n\varnothing)\exp\left(-\widetilde{\gamma_{mn}} \cdot z\right)$$

where:

$$\widetilde{\gamma_{mn}} = \sqrt{h^2_{mn} - \omega^2\mu\epsilon}$$

Annexure A6: VNA One Port/Two Port Calibration Procedure (40 GHz)

1. Switch on VNA.
2. Click on calibration.
3. Click on calibration wizard.
4. Set frequency.
5. Click on next.
6. Select on port.
7. Starting frequency range and stop frequency.
8. View calibration/kit.
9. Calibration response Port 1/Port 2.
10. Set the calibration kit to 85052B (3.5 mm).
11. Apply the open, short, and broadband at the load side.
12. First, connect open then choose male/female connector option. For antenna as one port, select female option.
13. Second, connect short to load, and similarly choose male/female.
14. Third, select broadband, and select female for antenna as one port.
15. Now remove the broadband and connect the antenna for the measurement.
16. Check S-parameters with the help of marker.
17. Data (for.csv) and Graphs (*.jpg*).
18. For two port device it will be female option.
19. Repeat above points 4–6.
20. Rest all same as was done for single port option.

Figure A6.1 is showing 40 GHz, vector network analyzer (VNA) for measuring DRAs-parameters.

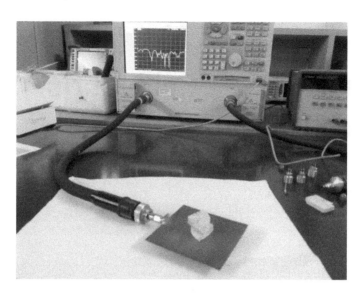

FIGURE A6.1   VNA for DRA measurements.

Annexure A7: Nano Dielectric Resonator Antenna for Biomedical Applications

1. Nanotechnology is currently in demand in medical sciences and communication. Nano DRA operates at terahertz frequency in spectrum. It is nondestructive and used in RF imaging. This technique also makes use of electromagnetic radiations. This technology can enable real time and high-resolution imaging. Nano DRA is also known as terahertz antenna used for imaging and can also be used for short range communications as it has large bandwidth for 5G applications.

2. Gaussian beam input creates nonlinear radiation. Graphene substrate has fascinating electrical and optical nonlinear properties. Graphene possess high plasmon sensitivity at THz frequency. Sample permittivity change at different frequencies can be utilized for biosensing. Here, third order harmonics are generated due to Gaussian input beams. Laser is used to excite nano-antenna via a silver waveguide to create surface plasma polaritons (SPPs). Graphene has strong light matter interaction, also it is a *third* order nonlinear response. The radiation at terahertz takes place. THz pulse generation takes place by ultra short femtosecond laser when incident on photoconductive antennas.

3. (THz = 10^{12} Hz) falls in between microwave band and infrared band, i.e., 100 GHz to 10 THz. Hence, at single fundamental frequency, different reflections from the same tissue layers can be achieved due to generation of harmonics. This has happened due to higher order mode generations. Tissue temperature rise becomes the main reason of absorption at THz regime. This is due to change in permittivity of specimen sample used. If these radiations are focused on healthy and unhealthy tissues, their absorbing properties and reflecting properties will be different because of water content. Reflected and absorbing signals will vary due to change in permittivity of layered tissues. This phenomenon is used for imaging. This phenomenon is quantified and compared with previously available data to predict the type of disease observed in the unhealthy tissue.

4. The terahertz frequency has the advantage of smaller wavelength. Hence, depth of signal penetration will be less as compared to microwave. However, the depth of penetration can further be varied with power input level. Terahertz frequency is a nonionizing and noninvasive frequency spectrum. THz imaging in tissues is carried out by making used change in

physiological structure and water content observations between normal and abnormal tissues. Change in power level may be used to capture images of layered tissue. This can have better resolution results with minimal investment as compared to MRI.

5. A terahertz device size shall be very small as compared to MRI instrumentation. Dielectric behaves like a conductor at high frequency and a conductor behaves like dielectric at high frequency. Biosafety is an important issue in sensing as the electron energy of THz wave is low. Biomedical technology (tissue exhibits reflection and absorption properties that change dramatically with tissue characteristics) is used for medical and dental imaging. Here, it is proved that dielectric response of the cell can reflect particular water dynamics by THz imaging.

6. Better image resolution is achieved by studying scattering parameters using strong near field applications. Waves ranging from 100 GHZ to 10 THz have unique excellent penetration ability in water content tissues and dielectric materials. Because of this, bioprocesses are studied at THz. Moreover, X-rays are not safe for human beings. THz waves are safe for biological detection and sensing. The morphological pattern is observed from the amplitude and phase information of reflected and absorbed e.m. signals. It can scan and provide information between abnormal and normal tissue.

7. A femtosecond pulse laser can produce THz pulse with a biased photoconductive antenna, then that is collected, collimated, and focused onto tissue. Reflections and absorptions are studied for high-resolution images. Terahertz (THz) waves are considered to be safe and contact-free technology for biological tissue pattern study and disease detection. Graphene with SiO_2 substrate is used in terahertz antennas. It is a THG (third harmonic generation) material. Terahertz radiation can pass through clothing, plastic, paper, cardboard, wood and ceramics. Terahertz imaging can be used to study the screening for weapons, explosives, and imaging of concealed objects.

8. Absorption can be measured using VNA with single port analysis and two port analysis. S_{11} is measured in two configurations, first with metal reflector and later placing the sample on the metal reflector. The difference of reflected power with these two readings will give us absorption coefficient. This measurement must be done in far field set up. This RF test can also be performed by connecting two ports: one to receive and other port as trans and keeping the target sample in between to measure reflections S_{11} and forward power S_{21} as absorption.

9. Nano DRA has wide bandwidth, hence is excellent for use in 5G applications. It has biomedical imaging nondestructive capability. It can be used for wireless security as well. Terahertz radiations can penetrate through clothing and dielectric materials but are reflected by metals. They are proximity coupled, and the feed used is silver nano waveguide. In nano DRA, the phenomenon of annihilation and creation of photons take place. Quantum electromagnetic fields are produced by current density.

The second quantized current density can be built out of the Dirac field of electrons and positrons while the free electromagnetic or photon field is built out of solutions to the wave equation with coefficients being operators, namely the creation and annihilation operators of the photons. The Dirac field also consists of free waves whose coefficients are operator fields in momentum space, namely the creation and annihilation fields of electrons and positrons. After computing the electromagnetic field produced by a second quantized current, the problem of determining quantum fluctuations in these fields in a given state of the electrons and positrons is deduced. The quantum electromagnetic field is produced by a quantum current density in optical antenna. The desired pattern of fluctuations or more generally, a desired pattern of higher order moments in space-time is obtained. Here, control input applied to the antenna will be in the form of a classical voltage/current source that will affect the Hamiltonian of the system of atoms that constitute the quantum antenna. The terahertz (THz) antenna can facilitate RF imaging, future security, wide band communications, sensing, actuation, and spectroscopy. Biological tissue absorbs and gets reflected due to different permittivity layers. They are captured for generating image. Thus, RF imaging concept is adapted with NDRA.

10. Mobile communication in 5G can operate at a sub-terahertz (sub-THz) band, i.e., (100–350) GHz, with data speed 20–100 Gbps. At THz frequency band, loss is more, and it can be 100 dB/km. Also, at 1 THz, atmospheric attenuation is relatively low, hence for 5G applications, the gain and bandwidth of the antennas should be high. Most of the research work available at THz in antenna are simulations based with CST Microwave Studio or Ansoft HFSS, and very few get converted into hardware.

11. Plasmonic nano-antenna's equivalent electrical length is much bigger as compared to its physical length. SPP waves in the plasmonic antenna have low traveling speed compared to that of free-space EM waves in microwave antennas. SPP waves can be tuned by nano-antenna material doping concentrations.

12. Developing nano-devices is still a challenge as the device dimensions are small. The miniaturization of a classical antenna to meet the size requirements of nano-devices will improve antenna efficiency. With advancement in manufacturing technology, miniaturization at THz antenna have now become a possibility. It is a fact that resonant frequency owing to frequency of 0.75 THz SPP wavelengths of an antenna wavelength is smaller by 10 times as compared to free space wavelength. Hence, radiated frequency is smaller as compared to input excitation frequency in plasmonic antenna, quantum antenna, and optical antennas.

13. Terahertz (THz) nano DRA operates at 0.1–10 THz, also known as photonic devices. Because THz radiation can pass through materials such as plastics, paper, and cloth and is nondestructive in nature, it is also non-ionizing. Hence it is considered safe for use in security and biomedical applications in high speed imaging (Figure A7.1).

FIGURE A7.1   Graphene oxide material used to design nano DRA.

14. The Purcell effect is generally described based on quantum electrody-
    namics as the effect of weak coupling from emitter to resonating cavity
    in terahertz nano-antennas. This way nanophotonics is differentiated with
    microwaves. The equivalent circuit model for nano-antenna shall be dif-
    ferent from microwave antenna. Any antenna at resonance is known as
    oscillator with radiating losses by taking into account radiation resistance.
    The frequency from 100 GHz to 10 THz is terahertz band where nano-
    antenna is different from optical and microwave antennas due to high
    absorption and scattering factors. These terahertz antennas are used for
    wireless robotic materials, in-body communication, on-chip communica-
    tion, and software-defined metamaterials.
15. Polymer DRA can be developed using THz frequency. 3D ink printers can
    be used to develop polymer DRA. Polymer DRA is a futuristic technol-
    ogy to be used for antenna designs. These polymer DRA can be used for
    biomedical applications at nanoscale. Wearable polymer DRA have huge
    potential in defense applications. Polymer DRA are made using flexible
    polymers and are an alternative to fabric-based embroidered antennas.
    Polymers are low-permittivity DRA used for higher frequency applications.
    These also fall in the flexible antenna category. Polymer antennas can be
    used as body sensors and can be developed with 3D inkjet printers. Polymer
    antennas are suitable for developing body area network. With the help of
    3D inkjet printing, prototyping of conformal DRA has become a possi-
    bility. For example, stretchable silver conductive paste is brush-painted on
    the Ninja Flex 3D-printed substrates. Graphene films have been used for
    nano-antenna designs. Polymer pipes filled with TiO_3 can be converted into
    polymer DRA, and they can be realized for new applications at nanoscales.
16. Nano DRA operates at the nanometer scale. Antenna at terahertz and opti-
    cal range can have wavelengths in order of atomic size of materials used
    in nano-antennas. When input excitation at optical range becomes equal to
    atomic size, resonance may occur. This resonance will be applicable to all
    the atoms subjected to the input excitations. Thus, this may be a phenom-
    enon of array of independent antennas formed by resonances created by

each atom. It will take the shape of a cloud formed by an array of antennas in resonance state. Hence summation of gain offered by each antenna will be available in this type of optical antennas. Also, the difference in wavelengths of input and atomic structure may offer little deviation in resonant frequency. This concept can be used to solve design problems of nano-antennas.

17. Nano-antenna with graphene layers placed above gold or silver patch can be used to convert wideband antennas to multiband antennas at nanoscale. They can be tuned to different bands by varying chemical potentials of graphene strips placed on nano DRA or nano-antenna.

18. **The retina as nano-antenna**: Retina cells within our eyes emit biophotons. Biophotons are biologically produced light particles, also known as photons (light particles). Biophotons in the human eye are generated due to oxidative metabolism. This is a chemical process in which oxygen is used to make energy from carbohydrates (sugars) through cell respiration. The layer above the retina has millions of light-sensitive receptors consisting of rods and cones. These receptors have pigment molecules, and chemical potential is developed by each molecule when light is incident on it. This triggers biological signals to be transmitted to the brain by absorbing photons. Hence, the retina absorbs light that is converted into electron chemical impulses and sent to the optical nerve, and then further to the brain.

A single particle of light is known as a photon. The biological generated impulses are called neurons. Photons have few electron-volts of energy. In the retina, rod cells amplify this small signal into measurable electrical response, known as neurons. Each rod is a single photon detector. In the retina, each cell can be visualized as a single quantum antenna. Hence, an array of quantum antennas is present at the retina. Each cell can detect light wavelength and frequency and delivers information. The brain collects this information to capture an image on the retina. Arrays of antennas are formed by the retina.

The eye retina has cones and rods, which acts as optical antenna to receive photons. The human **eye** has the ability to **receive signals** from about 430–750 THz, i.e., a full range of 320 THz as system bandwidth. Not only is it tuned to frequencies across this broad spectrum, it is also extremely directional. The other amazing thing about the human eye as an antenna system is that it has the ability, through its 100 million rods, to simultaneously look at the magnitude of a signal across a broad range. The eye has array of millions of nano-antennas formed by rods which can receive and transmit one photon, which matches to 430–750 THz, i.e., 320 Thz bandwidth.

19. Light is reflected by natural objects incident on the cornea of eye captured in the visible spectrum and reflected by a tissue. Light enters through the lens and is focused on the retina through the vitreous humor, which is a gel-like material consisting of 98% water. The retina has 126 million rods and cones known as photoreceptors. Rods have a wavelength of 496 nm. Cones have 420, 530, and 560 nm wavelengths. These photoreceptors can

absorb light in the frequency spectrum from 430 to 750 THz, also known as the visible spectrum. These photoreceptors have the capability to transduce light (photons) into electrical impulses (neurons). Photoreceptor cells have plasmonic membranes and vision pigment molecules called rhodopsin. These photons react with rhodopsin, the phenomena of electrochemistry takes place and electrical current of very small magnitude is generated in the form of neurons. Neurons can communicate in chemical form as well as in electrical form. There are networks of millions of optic nerves, which carry these neuron signals from the retinal surface to the brain. The brain does the processing part at a very fast rate to develop the image corresponding to a scene. Each photon receptor cell constitutes one pixel in the brain, which will enable development of a two dimensional image. These photons are collected by both the retinas and processed by the brain. These captured scenes are temporal and spatial. The phenomenon of real time is processed at a very fast rate in the brain.

20. Rods cells are responsible for creating black and white images and cones cells create color vision. Rod cells create photopic vision and cone cells create scotopic vision. Rods provide peripheral vision and cones provide central vision. The diameter of rod cells is 1–3 μm and the length of rod cells is 6–50 μm. Rods are in cylindrical shape and cones are of cone shape. The refraction index of eye tissue is 1.336 in the visible spectrum. Specific absorption rate (SAR) is measured as 2 W/kg on 10 gm tissue. Permittivity of rod tissue is 69 f/m and conductivity is 1.53 s/m.

21. Vision process: Use of bio-compatible electronic devices for prosthesis
    a. To develop an electronic device for possible replacement of dead photo-receptors in retinal layer.
    b. Compatibility of electronic device output with bipolar cell graded potential.
    c. Major part of image processing involves cornea, lens, vitreous humor, bipolar cells, ganglion cells, rods, cones (photoreceptors) with epithelium as absorber, optic nerves, and brain.
    d. Reflected light from natural objects falls on cornea and travels through vitreous humor and reaches to photoreceptors where it gets converted into electrical signals.
    e. These photoreceptors are formed by 126 million rods and cones in retina.
    f. Cones and rods have several layers of plasma membranes and photosensitive vision pigment molecules. They have the capability to absorb light by making use of rhodopsin and converting it into electrical signals.
    g. Rods are responsible for black and white vision and are also responsible for peripheral vision.
    h. Rods are located at the periphery of the retina.
    i. Cones are responsible for colored vision and located at the center of retina.
    j. Photoreceptors absorb and amplify incident light, and transduce it to a neural response for further communication to the brain through a network of optic nerves.

k. Image is developed from the scenes in the brain, i.e., photons get converted into pixels.

l. Photoreceptors are at an initial stage in the vision process.

m. These signals are progressively processed by bipolar cells as graded potential. At next stage, ganglion cells participate to convert the graded potential signals into gated potential signals. Now these signals can be termed as neural signals.

n. Each ganglion cell gated potential signal travels to the brain through optic nerves.

o. Neurons have the capability to communicate through chemical potential as well as through electrical potential.

p. Temporal and spatial phenomena on a real time basis take place in ten layers of the retina before developing an image in the brain.

q. The resolution of the eye cell depends on the number of photons received by quantum antennas in prosthesis.

r. An array of quantum antenna can be used as retinal implants.

s. Vision in the human retina is circularly polarized.

Index

Milton Keynes UK
Ingram Content Group UK Ltd.
UKHW031533071024
449327UK00005B/83